A TRIBBLE'S
GUIDE TO SPACE

Alan C. Tribble

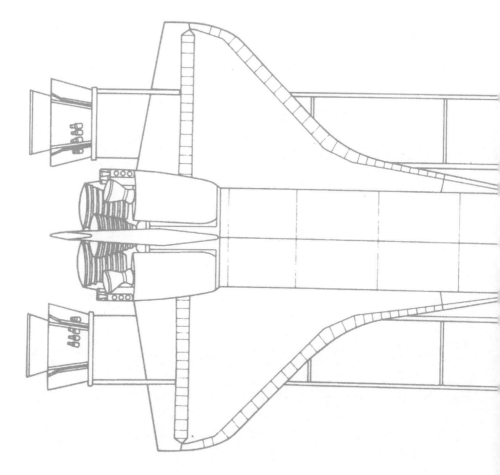

A TRIBBLE'S
GUIDE TO SPACE

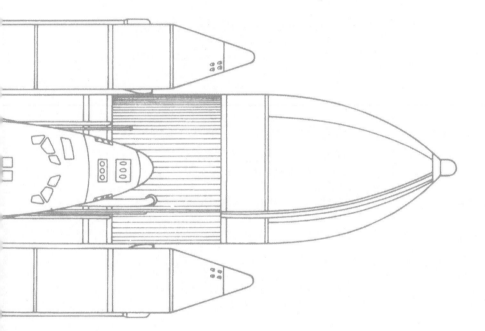

Princeton University Press • **Princeton and Oxford**

COPYRIGHT © 2000, PRINCETON UNIVERSITY PRESS
PUBLISHED BY PRINCETON UNIVERSITY PRESS,
41 WILLIAM STREET, PRINCETON, NEW JERSEY 08540
IN THE UNITED KINGDOM: PRINCETON UNIVERSITY PRESS,
3 MARKET PLACE, WOODSTOCK, OXFORDSHIRE OX20 1SY

LIBRARY OF CONGRESS CATALOGING-IN-PUBLICATION DATA

TRIBBLE, ALAN C., 1961–
A TRIBBLE'S GUIDE TO SPACE /
ALAN C. TRIBBLE
P.   CM.
INCLUDES BIBLIOGRAPHICAL REFERENCES AND INDEX.
ISBN 0-691-05059-7 (ALK. PAPER)
1. SPACE SCIENCES—POPULAR WORKS. I. TITLE.
QB500.25.T75 2000
500.5—dc21   00-036691

THIS BOOK HAS BEEN COMPOSED IN BERKELEY BOOK,
STAR TREK CLASSIC, AND STAR TREK TNG CRILLE

DESIGNED BY CARMINA ALVAREZ
COMPOSED BY EILEEN REILLY

THE PAPER USED IN THIS PUBLICATION MEETS
THE MINIMUM REQUIREMENTS OF
ANSI/NISO Z39.48-1992 (R1997) (*PERMANENCE OF PAPER*)

WWW.PUP.PRINCETON.EDU

PRINTED IN THE UNITED STATES OF AMERICA
1   3   5   7   9   10   8   6   4   2

CONTENTS

PREFACE

I was born three months after Alan Shepard's historic suborbital flight in May 1961. As such, I have always considered myself to be a member of the first generation of spacefarers from planet Earth. Some of my earliest memories are of watching the launch of the Gemini missions on television with my father and grandfather, who couldn't understand why it was so much fun to yell "blast off!" when the clock ticked down to zero. One month before I turned eight, I got to stay up past my bedtime to watch Neil Armstrong take "one small step for man, one giant leap for mankind." I stayed tuned in through the lull in the space program in the late 1970s, buoyed by regular doses of Carl Sagan's *Cosmos* television series and late nights in front of my small telescope. By then I was hooked. I never had any doubt that my life would be well spent if I could somehow find a way to take part in the great exploration of space. What other possible occupation could offer the opportunity to move beyond the routine things we worry about daily and lead to something truly awe inspiring?

Space has the possibility to unite mankind in spite of all of our differences. Every astronaut or cosmonaut who travels to space and back points out that from a few hundred miles in space, all of our problems, differences, and discontentments seem to shrink to insignificance. Within all of us is the feeling, the strong desire, to be part of something greater than ourselves. Why else would we be fascinated with television shows like *Star Trek*, or movies like *Star Wars*, *Apollo 13*, or *The Right Stuff*? Surely it is because they offer us the opportunity to devote our energies, passions, and enthusiasms to the pursuit of the great ideals. To paraphase the pilot's rhyme, surely we all have a desire "to tred the high untrespassed sanctity of space, reach out our hand, and touch the face of God."

During the past decade I have had the wonderful opportunity to be part of the space program. I had the chance to study space physics where it all started, at the University of Iowa, where Dr. James Van Allen and his team are still actively doing space physics research forty years after the launch of *Explorer 1*. While there, I had the chance to be part of a Space Shuttle mission and share in the glories of the past, and in the anticipation of the future. Later I had the opportunity to work for Rockwell International's aerospace operations in California. I will never forget the thrill of walking into the facility for the first time and realizing that this is the place that built the Apollo spacecraft and the Space Shuttle. I remember the pride echoed by my mentors who quietly reminded me, "We may not be the smartest people in the world, but we did go to the Moon."

As I started to build my career, I had the opportunity to work on the design of over a dozen spacecraft, from the Space Shuttle and International Space Station to the Global Positioning System (GPS) satellites, Mars landers, communications satellites, and military surveillance craft. Each new spacecraft showed a different use that we have found for space, uses that provide new and unique opportunities or sometimes only more effective and less costly solutions compared to ground-based approaches. I had the chance to watch the U.S. and Russian space programs switch from an atmosphere of competition to one of cooperation, and I even had the chance to work with Russian space scientists a time or two.

I like space, and I like to associate with people who share my enthusiasm, so a few years ago I started teaching courses on space. No matter where I go, I find that people everywhere are fascinated by space and want to learn more. Unlike last month's hit song or last winter's fashion trends, interest in space continues to grow over time. More people showed up to watch John Glenn blast off into orbit on his second mission in 1998 than did on his first in 1962. In the 1960s, when the Moon race was going strong, the television series *Star Trek* lasted only three years. In recent years, there have been three separate *Star Trek* spin-off shows on TV, plus several movies. Countless other space shows, from the fictional *Star Wars* to the biographical *October Sky* or the historical *Apollo 13*, continue to draw large audiences. Regardless of the time frame—past, present, or future—space is an interesting subject.

This book is intended to help those interested in space understand the basic "who, what, when, where, why, and how" about the space program as it stands at the dawn of the twenty-first century. I wrote the book in nontechnical terms to make it accessible to anyone who has an interest in space. The exploration of space has more potential to unite people from all cultures than any other endeavor. Commercial uses of space, which today are in their infancy, are literally poised to take off in the next few years. The past century was only a start. We journeyed to the Moon, sent probes to most of the other planets in our solar system, and figured out that many tasks, such as communications, are simply a lot cheaper to do up there. In the next century, who knows what may happen. Orbiting luxury hotels, spring-break cruises to the Moon and back, and manned explorations of other planets are all being studied in detail. It's going to be an exciting time, and you're invited to come along.

ACKNOWLEDGMENTS

As with most publishing projects, what starts as a simple idea in the mind of the author evolves into a much better final product with the help of many individuals. First and foremost, I am blessed with a wife, Beth, who supports my odd desire to spend time writing rather than doing those mundane things necessary to run a stable household; without the luxury of having the time to devote to this activity, the book would never have become a reality. My editor, Trevor Lipscombe, took painstaking efforts to guide the development of the manuscript in a direction the readers would enjoy. Many of the examples cited in the text are his. Numerous other reviewers, including James Van Allen, have made helpful suggestions to improve the readability of the final product. I thank all of them for their assistance.

A TRIBBLE'S
GUIDE TO SPACE

To Infinity and Beyond:
A Brief History of Space

Space. The final frontier.

When Buzz Lightyear heads for space, he sets a course to infinity and beyond. Fortunately for you and me, space is not *that* far away, and our travel plans don't have to be quite as ambitious to get us there.

Planet Earth has a radius of about 4000 miles. Let's imagine a sphere that encircles the Earth at a height of 50 miles above sea level. We'll call the region between the ground and the 50-mile altitude the Earth's atmosphere. Everything beyond 50 miles we'll call space. This means that space is closer than you think. If you're sitting in Chicago, you are actually closer to space than to Milwaukee, Wisconsin; and New York City is closer to space than to Philadelphia. There is really nothing magical about a 50-mile altitude. The transition from the Earth's atmosphere to space is not abrupt, so a rocket ship that travels through it would notice little difference as it passed from 49 miles to 51 miles. In the end, though, we need a reference point. Fifty miles is well above the altitude of Mount Everest, which is 5.5 miles high, or above cruising airliners that are 6 or 7 miles high. But it is also below the altitude of most orbiting spacecraft, so this artificial boundary is good enough.

Flight enthusiasts use terms such as the *troposphere, stratosphere, mesosphere*, and so forth to describe the various layers of the Earth's atmosphere. Likewise, we give different areas of space different names. The region between about 50 and 500-miles altitude is called *Low Earth Orbit*, or simply LEO. The upper boundary to LEO, like the 50-mile boundary to space, is rather artificial, but easy to remember because it is ten times the height of the lower boundary. Today, LEO is about as far as most of the spacecraft launched from Earth ever get. It is here where we find the largest payloads, like the Space Shuttle and the International Space Station.

Of course, a few spacecraft need to go higher and make it into what we call *High Earth Orbit* (HEO), the region between 500 miles and about 50,000 miles (which is about 12.5 Earth radii). Within this region lies a very special altitude at about 22,370 miles. Spacecraft placed here circle the Earth with the same speed at which the Earth rotates, making them appear to remain stationary above the Earth's surface. This is called the *synchronous orbit*, or more specifically geosynchronous or simply GEO. GEO is a very popular orbit for many spacecraft, especially for communications spacecraft that send and receive signals from specific points on the Earth's surface, and for spy satellites whose mission is to record what's going on at some specific location.

Above HEO, physicists have noticed some significant changes in the local environment. Most notably, above this point the Sun's magnetic field is stronger than the Earth's, causing the Sun, and not the Earth, to control what little atmosphere can be detected. Furthermore, once spacecraft get this far away from Earth, they rarely stay in orbit but move on to other parts of the solar system. Consequently, above HEO lies *interplanetary space*. Unless we choose to go into orbit around another planet, we would remain in interplanetary space until we are well past the orbit of Pluto. Relative to the size of our Earth, interplanetary space is vast. The distance to Pluto, the outermost planet, is over 3.6 billion miles, a distance so great that it takes light from the Sun over five hours to reach it.

The ancient Greeks were the first to try to determine the size of the solar system using simple geometry. Aristarchus, Eratothenes, and Hipparchus all made various estimates of the Earth-Sun distance. Although their original measurements were too small, they showed clearly that the solar system was hundreds of times larger than the Earth itself. The great distances between planets had been appreciated for centuries, but Galileo Galilei, over three hundred years ago, was the first to see firsthand just how large interplanetary space is. With his new invention, the telescope, he even saw that Jupiter had its own moons circling it. As the years passed and observations of the moons of Jupiter continued, observers developed the ability to predict when the moons would emerge from behind the great planet.

These early astronomers noticed, however, that sometimes their predictions were off. The moons of Jupiter would appear several minutes early, or they would appear several minutes late. They finally deduced that the appearance of the moons occurred later than expected whenever the Earth and Jupiter were on opposite sides of the Sun (fig. 1). The predictions were accurate, but the early astronomers did not realize that it takes light several minutes to travel the great distances involved. When the Earth and Jupiter were on the same side of the Sun, the light would arrive early; when they were on opposite sides of the Sun, the light would arrive late. Olaus Roemer, a Danish astronomer, was the first to realize that this delay was due to the finite amount of time it takes light to cross these great interplanetary distances. He used this insight in 1675 to

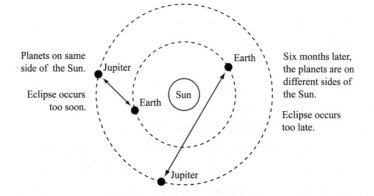

Figure I. The moons of Jupiter went into eclipse sooner than expected when Earth and Jupiter were close. The eclipses occurred later than expected when Earth and Jupiter were far away. The time difference is the time it takes light to travel the extra distance.

make the first crude measurement of the speed of light, 186,000 miles per second.

As we start to explore beyond our own solar system, somewhere past Pluto we will expect to find a point where the Sun's magnetic field has weakened so much that the small magnetic field in the interstellar medium becomes stronger. This point will mark the beginning of *interstellar space*. As big as interplanetary space is, it is still dwarfed by the distances to the stars. The first estimates of these distances were made using *parallax* measurements. Parallax is an easy concept to understand. Hold your thumb up in front of your face. Close your left eye and look at your thumb relative to the background with your right eye. Now close your right eye, open the left, and note how your thumb "appears" to have moved relative to the background. That shift is called a parallax. Parallax measurements of the stars are made by allowing the Earth to take the place of your eyes, and a close-by star to take the place of your thumb. We simply take a picture of the star in question from the Earth, then wait six months and take another picture when the Earth is on the opposite

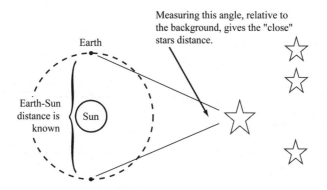

Measuring this angle, relative to the background, gives the "close" stars distance.

Earth

Earth-Sun distance is known

Sun

Figure 2. By measuring the angular shift in the position of a close star, called its parallax, astronomers can measure the distance to the stars since we know the Earth-Sun distance.

side of the Sun (fig. 2). By comparing pictures taken when the Earth is in two different locations, we can determine the parallax to some of the stars closest to the Earth. F. W. Bessel, a German astronomer, first succeeded doing this in 1838 and found that the stars were over a million million miles away, a distance so great that it takes light from other stars years, or centuries, to reach us. This realization opened up the size of the Universe immensely.

As we continue our outward exploration, we will pass other star systems with planets of their own, but we will eventually come to the edge of our own Milky Way galaxy and move into *intergalactic space*. Even if we make it this far, we still have a long way to go. The Andromeda galaxy, for example, is over 10 billion billion miles away, and it takes light over 2 million years to reach us from this great distance. Because of these huge distances and our limited history of sending machines into space, the Universe is still too large for us to explore in person (fig. 3).

Humans have dreamed of exploring space for countless centuries. According to Greek mythology, the first two men to become airborne were Daedalus and his son, Icarus, two prisoners on the island of

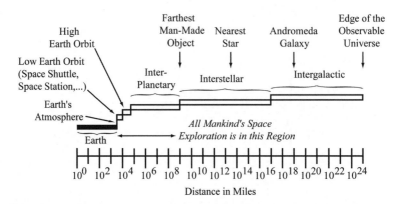

Figure 3. The Universe is too large for mankind to be able to explore in person.

Crete. Impatient while waiting for King Minos to pardon them, Daedalus fabricated two pairs of wings by gluing feathers to a wooden frame with wax. With their wings attached, father and son were soon airborne. Daedalus warned his teenage son not to fly too near the Sun lest its heat melt the wax, but the young Icarus became so absorbed in his new ability to fly that he did not heed his father's words. He flew too near the Sun, the wax melted, his feathers fell off, and he dropped into the sea, to his untimely death.

For the next several centuries, wings preoccupied the imaginations of those who wished to fly like the birds. Then, in 1766, Henry Cavendish discovered what he called "inflammable air"—hydrogen—and showed that it was much lighter than the surrounding atmosphere. Knowledge of this new discovery inspired French paper manufacturer Joseph Montgolfier to begin experimenting with lighter-than-air machines. In 1782, with his brother Etienne, he succeeded in flying small unmanned paper and cloth balloons filled with hot air. By November 21, 1783, they were ready to try a manned flight in a large balloon made of cotton and paper, coated with alum as a means of fireproofing. Cords sewn into the fabric carried a wicker gallery at the base. In between was a large brazier

To Infinity and Beyond

that contained a fire that could be fed with straw. Unsurprisingly, the Montgolfier brothers were reluctant to fly in a paper balloon with a roaring fire inside of it, so they arranged for others to serve as pilots: the Marquis D'Arlandes and Pilatre de Rozier. Although the balloon burned in a few places, it survived long enough to give the men a twenty-five-minute flight over the rooftops of Paris. They reached a height of 3000 feet before landing in a park outside the city. Over two hundred years later, on March 20, 1999, Bertrand Piccard and Brian Jones became the first aviators to circle the globe nonstop in a hot-air balloon. Their trip lasted 19 days, 1 hour, and 49 minutes.

As nice as balloons were for getting an aerial view of things, they left you completely at the mercy of the local winds. Balloonists learned to control their altitude, but they could travel only in the direction in which the wind was blowing. Birds, on the other hand, could fly hither and yon as they pleased. Consequently, it was only a matter of time before interest in flying shifted back to winged machines. By the end of the twentieth century, many people had successfully piloted winged gliders, but, lacking a power source, they too were at the whim of the winds.

Two brothers from Ohio, Orville and Wilbur Wright, decided to become serious "students of the flying problem," as they liked to call themselves. Approaching the problem in a scientific manner, the brothers realized that in order to develop a functioning airplane they would have to solve three problems: (1) lift, provided by the wings; (2) power, provided by an engine; and (3) control. After experimenting with wing shapes (to generate lift), propeller shapes (to power the airplane), and different ways of twisting the wings (to control the direction of flight), the brothers were ready for a test. On December 17, 1903, at Kitty Hawk, North Carolina, they succeeded in proving the concept of powered flight with a 200-foot sprint lasting a mere fifteen seconds. But by 1910, they were flying for more than 20 miles at a time and remained airborne for over an hour and a half. In 1947, less than forty years later, Chuck Yeager was already able to fly his Bell X-1 aircraft faster than the speed of sound. Try as they might, however, airplanes could not fly more than a few miles high because the air at those altitudes is too thin, essentially leaving nothing for the plane to fly through. To get to space, it was clear what was needed: a means of generating lift without relying on wings or the air pushing up underneath them.

In 1865, Jules Verne had proposed that we get to the Moon by shooting people out of a large cannon. George Melies, the writer, director, and actor of the earliest science fiction film of all time, *Le voyage dans la lune* (1902), chose this method of travel as well. However, it was rockets, which could carry their own fuel and burn it continuously along their journey, that were destined to carry men into space. The Chinese had already been using rockets as early as the second century B.C., primarily for entertainment in the form of fireworks (fig. 4). Over the next two thousand years or so, rockets eventually worked their way into use as a means of warfare as well. "The rockets' red glare" was already noted by Sir Francis Scott Key during the War of 1812 and was immortalized in his poem, "The Star Spangled Banner."

As with most technological developments, the greatest advances have occurred in the past one hundred years. Near the dawn of the twentieth century, a Russian schoolteacher, Konstantin Tsiolkovsky, showed that a rocket—powered by liquid fuel—was the only practical way to get out of the atmosphere and into space. In 1919, an American student named Robert Goddard proposed that a rocket could be flown to the Moon. The public rapidly dubbed him an eccentric. But their reaction did not stop him from building and

Figure 4. The Chinese developed the first rockets, in the form of fireworks, over two thousand years ago. (Courtesy NASA)

launching the first liquid-fueled rocket from his aunt's farm in March 1926. It reached the amazing height of 150 feet.

Because of the reaction to his 1919 paper, Goddard religiously avoided publicity for the rest of his life. He even refused to aid the American Interplanetary Society's attempts to publicize his work. Members of the society visited Germany in 1931 and made contact with the German Rocket Society, whose membership included Hermann Oberth, the author of *The Rocket into Interplanetary Space* (1923). A German teenager named Wernher von Braun had read the book in 1925, and he was so intrigued by it that he had strapped some rockets onto his little red wagon and propelled it through the streets of his hometown before the fireworks exploded and the annoyed villagers summoned his parents. Von Braun had joined Oberth by 1930 in trying to develop functioning rockets, and the visiting Americans were very impressed with what they saw. Still, they were unable to stir up sufficient interest at home to get a research program up and running. The Germans did not have this

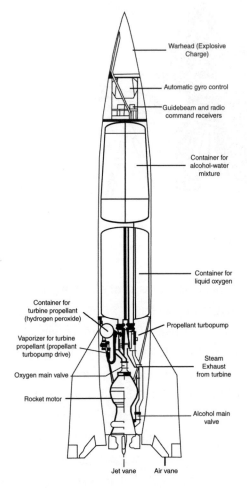

Warhead (Explosive Charge)

Automatic gyro control

Guidebeam and radio command receivers

Container for alcohol-water mixture

Container for liquid oxygen

Container for turbine propellant (hydrogen peroxide)

Vaporizer for turbine propellant (propellant turbopump drive)

Oxygen main valve

Rocket motor

Propellant turbopump

Steam Exhaust from turbine

Alcohol main valve

Jet vane   Air vane

German V–2 (A–4) Missile

Figure 5. The V-2 rocket, developed by Germany late in World War II, was the predecessor of America's first launch vehicles. (Courtesy NASA)

problem and were flying liquid-propellant rockets by 1934. By 1942, von Braun was leading the German rocket program and stayed at its helm during the remaining war years.

The most famous of the German rocket designs was the vengeance weapon number two, or V-2 (fig. 5), the first long-range ballistic missile, capable of flying from Germany to England. Germany was so protective of this secret that Hitler ordered the execution of von Braun and his team in the closing days of the war. However, von Braun's brother managed to contact the American forces, who quickly took custody of the team just before the German SS could enforce Hitler's orders. General Dwight D. Eisenhower, aware of the rocket's potential for warfare, allowed the German team to continue with a final series of tests after the war. A Russian colonel, Sergei Korolev, was allowed to observe the last few tests flights. Ten years later, Korolev would be head of the Soviets' rocket program.

In 1946 von Braun's team was transported to White Sands, New Mexico, in the American desert, where it continued its experimentation in rocket design. In 1950 the team moved to Huntsville, Alabama, and eventually became part of NASA. Finally, in 1957, a rocket was developed that could carry a payload into space. Unfortunately for von Braun and his colleagues, this feat was accomplished by the Soviet Union, which launched the first satellite, *Sputnik*, on October 4, 1957.

By 1958, only a year later, von Braun and his team were ready to launch the first U.S. satellite, *Explorer 1*. The group also played an integral part in launching America's first astronaut, Alan Shepard, into space in 1961. Unlike the shy Goddard, von Braun was able to survive the world war and continue his research largely due to his outgoing nature. On his desk, von Braun kept a plaque that advised: "Late to bed. Early to rise. Work like hell, and advertise!" Homer Hickam's recent novel, *Rocket Boys*, and the screen adaptation, *October Sky*, clearly indicate how popular von Braun was in the United States in the late 1950s.

Besides the glamour associated with the "rocket scientists" who were busy developing rockets, there was the challenge of keeping a crew member alive once the rocket was out in space. The balloonists had known about this problem for many years. When researchers suffocated or froze to death in the bitter cold during the early balloon flights, it quickly became obvious that some means of protecting the crew must be found. This had been achieved by November 1935, when the altitude record of 73,395 feet was set by two U.S. Army "aeronauts" in a balloon made of rubberized fabric.

By the late 1950s, high-altitude balloon flights were conducted under program names such as the air force's "Man-High" or the navy's "Strato-Lab." When Joe Kittenger bailed out of his open gondola as part of the Man-High project at an altitude of over 100,000 feet, his first reaction was to think that something had gone terribly wrong. In normal parachute jumps, you feel the force of the wind buffeting you relentlessly, but he felt nothing. He described the feeling as "like being suspended in space." For a fraction of a second, he wondered if the scientists had made a mistake in their calculations and he had actually flown high enough to be above the Earth's gravitational pull. However, when he rolled onto his back he saw the balloon rapidly disappearing above him as he fell toward Earth. After only a few seconds, the atmospheric density increased enough so that he could feel the wind rush by him. In May 1961, the Strato-Lab V balloon conducted a test of the Mercury space suits by carrying two men to an altitude of 115,000 feet in an open gondola. That altitude is above 99 percent of the Earth's atmosphere. Balloon tests such as these, and aircraft test programs such as the X-15, sometimes lay claim to having sent the first astronauts into "space" by flying high enough to see the curvature of the Earth.

As valid as those claims may be, the glory went to an entirely different group of people. The United States selected its first astronauts in 1959. At about the same time, the Soviet Union selected its first cosmonauts. The selection process was classified at the time, so little was known about the candidates until the "winners" were

Figure 6. The original Mercury astronauts. Back, left to right: Alan Shepard, Walter Schirra, John Glenn. Front, left to right: Virgil "Gus" Grissom, Scott Carpenter, Donald "Deke" Slayton, Gordon Cooper. (Courtesy NASA)

announced to the public. The original Mercury astronauts were chosen from a pool of five hundred candidates, all of whom were military pilots. The original selection criteria were quite brief: experience in flying jet aircraft, preferably as a test pilot; an engineering background; and a height of less than 5 feet 11 inches. The height requirement was an acknowledgment of the fact that real estate on a spacecraft is very expensive, and smaller astronauts, who can fit into smaller spacecraft, help reduce the cost. As shown in the movie *The Right Stuff*, a rather sophisticated series of tests eventually reduced the pool down to the final seven (fig. 6). Half a world away, the original cosmonauts were military test pilots as well.

The Soviet pilot Yuri Gagarin won the honors as the first man in space after a one-orbit flight in April 1961. Unknown to the public at

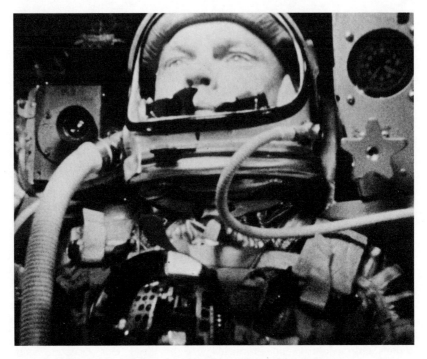

Figure 7. John Glenn, shown here in the cockpit of his Mercury capsule, was the first American to orbit the Earth in 1962. (Courtesy NASA)

the time, Gagarin actually bailed out of his capsule before it landed, parachuting the last two miles back down. His arrival on Earth was witnessed by an old woman who asked, "Have you come from outer space?" He replied, "Yes. Would you believe it, I certainly have." Then, to calm her fears, he quickly added, "But don't be alarmed, I'm a Soviet." Ironically, Gagarin was killed in a plane crash in 1968.

In order to keep up with the Soviets, several Americans flew into space in quick succession. Alan Shepard became the first American in space after a brief fifteen-minute suborbital flight in May 1961. John Glenn (fig. 7) became the first American to orbit the Earth in 1962. His great popularity with the public may have made it difficult for NASA to consider sending him back into space again. He left NASA in 1965 to pursue a business career and then was elected to

Figure 8. Jim Lovell, shown here in the cockpit of the *Gemini 12* capsule, was chosen in the second astronaut selection in 1962. He also flew on *Gemini 7*, *Gemini 12*, *Apollo 8*, and the ill-fated *Apollo 13* mission. (Courtesy NASA)

the Senate from his home state of Ohio in 1974. Thirty-six years after his first mission, he returned to space in 1998, the oldest person ever to do so, on the Shuttle. In 1999, NASA renamed its research facility in Ohio, formerly the Lewis Research Center, the Glenn Research Center.

After President Kennedy challenged the nation to "put a man on the Moon and return him safely to Earth," it was obvious that more than seven astronauts would be needed. The second group of nine astronauts, referred to as the "new nine," was selected in September 1962 and included Neil Armstrong and Jim Lovell.

Jim Lovell (fig. 8) began his space-flight career with a record-setting fourteen-day trip in *Gemini 7*. (That's fourteen days confined to one chair, with another astronaut, Frank Borman, sitting only 2 feet away.) He flew again on *Gemini 12*, then was part of the first crew to visit the Moon on *Apollo 8*. While he and his colleagues Borman and Bill Anders orbited the Moon on Christmas Eve 1968, television viewers across the United States watched as the astronauts read from the Bible's Book of Genesis. Later, Lovell was the commander of *Apollo 13* on a return flight to the Moon, but an explosion turned the "routine" flight into one of the most heroic rescue missions ever recorded. Twenty-five years later, director Ron Howard immortalized the event in one of the most detailed movies of the Apollo program ever made, entitled simply *Apollo 13*.

Another famous astronaut, Neil Armstrong (fig. 9), flew on *Gemini 8* in 1966 and was the first man to walk on the Moon as part of *Apollo 11* in July 1969. As he did so, he uttered the memorable, "That's one small step for man, one giant leap for mankind."

The third group of fourteen pilot astronauts, selected in October 1963, included *Apollo 11* Lunar Module pilot Buzz Aldrin and *Apollo 11* Command Module pilot Michael Collins. Aldrin seemed destined for spaceflight from the beginning. His mother's name was Marion Moon, and his father was a student of rocket pioneer Robert Goddard. He earned a doctorate in astronautics from the Massachusetts Institute of Technology, where he devised rendezvous techniques used on virtually all NASA missions. After walking in space on *Gemini 12*, he and Neil Armstrong drew the largest worldwide television audience ever during the historic *Apollo 11* event. Interestingly, all of the *Apollo 11* photographs of an astronaut on the Moon are of Aldrin. Armstrong carried the only camera and never got in the picture himself, except as a reflection in Aldrin's helmet (plate 1).

As the Gemini program neared completion, a group of six scientist-astronauts was chosen in 1965, including Harrison Schmitt, the only

Figure 9. Neil Armstrong, shown here in the Lunar Module, was the first man to walk on the Moon as part of *Apollo 11.* He also flew on *Gemini 8.* (Courtesy NASA)

scientist to walk on the Moon. This was followed by nineteen additional pilot-astronauts in 1966 and eleven more scientist-astronauts in 1967. With the Apollo program's days clearly numbered, the final selection of the decade was a group of seven astronauts selected for the Air Force Manned Orbiting Laboratory. Among these was Major Robert Lawrence, Jr., the first black man selected for astronaut training. He was killed on December 7, 1967, in the crash of his F-104 fighter during a training exercise. Although he never flew the magical 50 miles up to earn his astronaut wings, his name was added to the astronauts' memorial at Kennedy Space Center thirty years after his death in official recognition of his astronaut status. After the air force canceled the Manned Orbiting Laboratory program in 1969, the remaining men in the program joined the NASA astronaut corps.

A big dry spell occurred in the U.S. space program until 1978, when the Space Shuttle astronaut selection process got under way. Thirty-five new astronauts, including the first women and African Americans, were selected. Sally Ride was a Ph.D. candidate doing research in astrophysics when she saw the call for astronauts in the Stanford University newspaper. She became the first American woman in space in 1983 on flight STS-7, then flew again on STS-41G in 1984. She was preparing for her third mission during the *Challenger* disaster and was appointed to the presidential commission investigating the accident. In 1987 she left NASA and is now director of the California Space Institute.

Astronaut selection has proceeded since 1978 at roughly two- to four-year intervals. A total of 318 candidates have been selected in seventeen groups from 1959 through 1999. Today, about 120 astronauts are on active duty with NASA.

Unlike in the early days of space travel, a time when the only route to space was by being a military test pilot, most space agencies today have opened up the astronaut selection process to civilians. The most basic requirement is a science or engineering background—generally a four-year college degree—and experience. Candidates must also pass a physical exam, which will verify general good health, and meet a few other specific criteria. Finding your glasses in zero gravity ("zero-g") isn't easy, so uncorrected vision better than 20/100 is a requirement. Blood pressure cannot be too high, since an astronaut must be able to withstand the pressures of a launch and the effects of zero-g. A height requirement insures that you're not too tall to fit into the spacecraft or too short to reach the controls. A psychological exam is also required. Those who are unable to remain cooped up in a small space that is strapped on top of tons of high explosives need not apply. Astronauts must also like to work as part of a team and be able to get along with co-workers. After one hundred days in space onboard a small space station, every little habit your colleagues have may just become too annoying to deal with. Cosmonauts who have had extended stays on the Mir Space Station typically need about a year before they develop the desire to talk to one another again. It's not that they're mad at each other, it's because they have already been exposed to each and every story the other person has to tell his or her captive audience.

Today, NASA accepts applications from U.S. citizens for pilot astronauts, mission specialist astronauts, and payload specialists. Pilot astronauts, as the name suggests, serve as either the commander or pilot of the Space Shuttle. The commander (or captain) is responsible for the vehicle, the crew, the mission's success, and safety. The pilot is second in command (the executive officer) and assists the commander in controlling and operating the shuttle. To be selected as a pilot astronaut candidate, an applicant must meet the general requirements plus have completed at least one thousand hours of flying time in jet aircraft, preferably as a test pilot.

Mission specialist astronauts are responsible for coordinating the onboard operations of the Shuttle and for conducting experiment and payload activities. They are required to have a detailed understanding of its systems and of the objectives for each of the experiments conducted during a mission. Mission specialists perform spacewalks, or extravehicular activity (EVA), and handle many payload functions. They must have the same basic qualifications as pilot astronauts except for the flying-time requirement.

Payload specialists are not professional astronauts and, consequently, do not have to be U.S. citizens. They are specialists in the physical or life sciences, or individuals skilled at operating some unique payload equipment. Payload specialists are chosen by the payload sponsor or the paying customer. For NASA-sponsored projects, the specialists are nominated by the group providing the payload and approved by NASA. They must also meet strict NASA health and physical fitness standards. John Glenn's second trip into space was as a payload specialist.

Once selected for training, astronaut candidates attend a year-long "astronaut candidate school" at the NASA Johnson Space Center (JSC) in Houston, Texas, in order to qualify for spaceflight. All astronauts attend courses in aircraft and spacecraft safety and take a full range of science and technical courses as well. Survival training is a necessary part of this experience, as an emergency water landing is a very real possibility. The basic training is rounded out with knowledge of the Shuttle's systems, orbiter habitability, housekeeping and maintenance, waste management, and extravehicular activity. To prepare for EVAs, all astronauts must become SCUBA qualified.

All astronaut candidates learn to function in weightlessness with the help of the "vomit comet" trainer aircraft, so named because of a side effect it produces on many first time travelers, i.e., motion sickness. Additional weightless training can be simulated in the "neutral buoyancy" water tank at JSC (fig. 10). Pilot astronauts train in T-38 jet aircraft and a modified Grumman Gulfstream II. Although the astronaut corps considers candidates to be true astronauts, they don't become "official" astronauts until completion of the basic course. Advanced training follows the year-long basic course. Then, once they are selected for a mission, astronauts enter an additional period of mission-specific training for seven to twelve months before launch.

To date, all visitors to space have gone on either U.S. or Soviet (now Russian) launch vehicles. However, many countries have had the opportunity to contribute candidates through either the American or Russian space programs. Citizens of Afghanistan, Austria, Bul-

Figure 10. A large swimming pool, called the neutral buoyancy tank, at the NASA Johnson Space Center is used to help the astronauts experience what it's like to work in weightlessness. (Courtesy NASA)

garia, Canada, Cuba, Czechoslovakia, Germany, Hungary, India, Mexico, Mongolia, Saudi Arabia, the United Kingdom, and Vietnam have all visited space, and the participation of other countries is being planned. The National Space Development Agency of Japan, for example, has advertised for astronauts on three occasions and has recruited five candidates, all of whom are now in training in the United States. The European Space Agency (ESA) also selects and trains astronauts for missions with either the United States or Russia. China has a robust space program underway, too, and hopes to launch its astronauts on its own rockets before the end of 2000.

Those interested in becoming an astronaut can take a few key steps to increase their odds. As Colonel Charlie Bolden, former astronaut and deputy commandant of midshipmen at the U.S. Naval Academy, says, "Start with the basics and get them down first. You can't do anything without math and science." In addition to having a strong math and science background, candidates must also be able to work as part of a team; understand and appreciate their ethnic, cultural, and American history; and maintain a grasp on current events. It is interesting to note that about two-thirds of the former and present U.S. astronauts were active in scouting in their youth.

Astronauts are always in high demand by the public, so they must be able to be comfortable in the limelight. NASA routinely receives requests that astronauts share their knowledge with various groups, from schools to retirement centers. Communities that were once the home of these space travelers, or hope to be the home of future ones, proudly advertise this fact to visitors by naming streets, parks, or community centers after them. Within a few hours' drive of my house, you can find the "Loren Shriver Community Center" in Peyton, Iowa, a thriving metropolis of 350 people that is proudly named after the Air Force Academy graduate and Shuttle astronaut. Even Riverside, Iowa (population 826), proudly announces that on March 22, 2228, it will become the birthplace of Starship Captain James T. Kirk.

Good grades in science and technology are a necessity. Good communication skills, and even knowledge of a second language, are also important. There's nothing unique about these qualifications: they are basic, commonsense principles that prepare you to be more successful in any line of work. This is fortunate, for the odds of a person actually being selected are very slim. Note that of the five hundred applicants screened for the original selection in 1959, only seven were chosen. When NASA sent out the announcement for the first selection of Shuttle astronauts in 1978, over eight thousand people applied for only thirty five slots. Typically, the odds of being selected—provided you meet the selection criteria and actually go to the trouble of applying—are about 200 to 1. More recently, an average of about four thousand applicants have responded to NASA's announcement every two years. Of those four thousand an average of 118 are asked to go to the Johnson Space Center for a full week of interviews, medical examinations, and tests. Of those, about twenty are offered the opportunity to become astronaut candidates. Those that make it are offered long work hours and relatively low pay. It is not unusual for candidates to arrive at work by 7:30 A.M. and not get home until after 11:00 P.M. Astronaut pay begins on the government scale GS-11 at about $40,000 and tops off at about $80,000.

If we fight the odds and are fortunate enough to complete astronaut candidate school we will be ready for our first spaceflight. As we know, the first part of a mission is the blast-off, the start of an unforgettable journey into space. Once in space, we must know how to operate the systems on the Shuttle and oversee various experiments. To be up to the task, we need to understand how one actually gets into orbit, and what the spacecraft has to do once it gets there. Let's proceed to learn a little about what the astronauts must know—the subject of astronautics.

Beam Me Up:
Getting to Space

Standard orbit, Mr. Sulu!

The words roll off Captain Kirk's tongue with ease, for indeed orbiting a planet must be an everyday occurrence for the crew of the Starship *Enterprise*. Mr. Sulu pushes a few buttons on his console, and before you know it a planet is whizzing by underneath. Conceptually, we think of an orbit as the motion of a spacecraft relative to a planet, but what really is an orbit, and what would make an orbit "standard"? These problems have been studied for over three hundred years and will be studied for two hundred more before the *Enterprise* can become a reality. Still, today's space-mission designers would say that there is no such thing as a "standard" orbit because often the biggest problem that space travelers face is actually getting into space. Each space mission is a one-of-a-kind event, so a unique orbit must be determined for every spacecraft. As we saw in the previous chapter, space is not too far away; the Space Shuttle, for example, orbits at about the same distance above the surface of the Earth as New York City is from Washington, D.C. However, getting there—and staying there—is a formidable problem. To understand how spacecraft get into orbit, we must first understand what an orbit is.

To solve this problem, spacecraft engineers turn to the laws of physics, which can be used to develop the *equations of motion*— mathematical relationships that allow us to predict the future motion of a spacecraft given its current conditions. They link together position, velocity, and acceleration. To understand how the equations of motion can be used to define spacecraft orbits, let's go back about four hundred years, to the time of Johannes Kepler.

In the latter part of the 1500s, European scientists began to accept the fact that the Sun, not the Earth, was the center of the solar system. (The ancient Greeks had already proposed a geocentric theory about two thousand years earlier, but somehow this idea had gotten lost.) Renaissance astronomers believed that the planets moved in great circular orbits around the Sun, just as the Moon moved around the Earth. If they knew the position of a planet in the night sky at any given time, they could also predict the planet's future positions. Such predictions of the positions of the planets, the Moon, and the stars were of great interest because it allowed scientists to forecast eclipses, seafarers to navigate ships, and thus indirectly improved the general quality of life.

By the dawn of the seventeenth century, Tycho Brahe (1546–1601) and other scientists were already compiling accurate measurements of the positions of the stars and planets. Brahe was a curious character. He had an artificial nose made of copper, placed on his face after the original item was cut off in a duel. But he met his demise in another way. During a banquet in 1601 he became sick and suffered from fever and attacks of giddiness for many days. He died on October 24 due to what was diagnosed as urinary poisoning. However, at the end of the twentieth century, modern nuclear microprobing devices were used to analyze a few of Brahe's hairs, re-

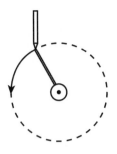

Figure 11. A circle can easily be drawn with a loop of string, a thumbtack, and a pencil.

trieved from his tomb, to search for lead or mercury. The results indicate that he may have been poisoned on the day before he died, perhaps in a failed attempt to cure himself.

Brahe was primarily an observer, not a mathematician, and it was left to his assistant, Johannes Kepler (1571–1630), to interpret his data. He discovered that the motion of the planets could be explained only if their orbits were ellipses, not circles.

To understand the difference, we can perform a simple exercise. Take a piece of cardboard, a loop of string, a thumbtack, and a pencil. Stick the thumbtack in the cardboard and loop the string around it. Attach the pencil to the other end of the string, and move it outward until the string is taut. By keeping the tension in the string constant, you can move the pencil in a loop and draw a perfect circle (fig. 11). By reasoning that the Sun (thumbtack) would exert an attractive force (string) on a planet (pencil), it must have seemed perfectly logical to Kepler that the planet would be attracted to the center of the Sun and the motion would be circular. Besides, the heavens were the realm of God and the angels, and so the circle, long regarded as the perfect shape, must be the path followed by the planets. To add a religious interpretation, as was common in his time, Kepler explained that the planets were pushed around on their orbits by angels, and that this action would produce a musical note—the

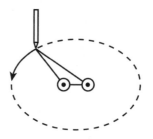

Figure 12. To draw an ellipse, we simply add a second thumbtack to our circle-drawing gear.

music of the heavens. However, Kepler's more accurate measurements of the position of the planets showed that their motion was not circular but elliptical. To draw an ellipse using the same equipment, we simply add a second thumbtack a short distance from the first. Loop the string over both thumbtacks, and, keeping the string tight, your pencil can now be used to draw an ellipse (fig. 12). The locations of the two thumbtacks are called the *focal points*, or foci, of the ellipse. A circle is really a special kind of ellipse, one whose focal points overlap.

Kepler's observation of the elliptical motion of planets around the Sun is true of all orbits in general. That is, a spacecraft orbiting around a planet will move in an ellipse, with the larger mass (the planet) being at one focal point and nothing of significance at the other. The "orbit" of a spacecraft around a planet is described by the size and shape of the ellipse (fig. 13) and its orientation relative to the planet's equator (fig. 14). The spacecraft's point of closest approach to the planet is called *perigee*, and the farthest point *apogee*. Depending on what the spacecraft is trying to do, its orbit may be very elliptical, so that the difference between apogee and perigee is great. At other times, orbits are so close to being a perfect circle that the small difference between perigee and apogee is ignored, and the orbit is simply defined in terms of an *orbital altitude*—its height above the surface of the planet. The angle of orientation of the ellipse

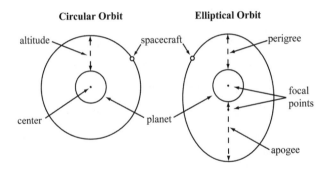

Circular Orbit   **Elliptical Orbit**

Figure 13. An elliptical orbit is defined by its point of closest approach, perigee, and its point of farthest retreat, apogee. If perigee and apogee are about the same, the orbit is approximately circular and can simply be described in terms of the orbital altitude.

relative to the planet's equator is called the *inclination* of the orbit. Kepler was the first to understand what an orbit is, but he did not know why it had to be elliptical—and he also did not know how to put something in orbit around the Earth. That breakthrough came later, with Isaac Newton.

For about seventy years after Kepler's first law of planetary motion—that planets move in ellipses with the Sun as one focus—the scientific community struggled to understand it. In the 1680s, after years of study, Isaac Newton (1642–1727) found that three fundamental assumptions, his three *laws of motion*, would enable him to understand how to get a spacecraft into orbit. Newton's three laws are as follows:

- An object that is either at rest, or in motion with constant velocity, will remain at rest, or in motion with constant velocity, unless acted on by an external force.
- Force is the rate of change of momentum, which is also the product of mass and acceleration.
- For every force (action) there is an equal and opposite force (reaction).

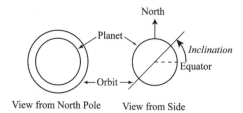

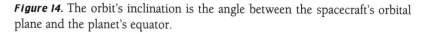

Figure 14. The orbit's inclination is the angle between the spacecraft's orbital plane and the planet's equator.

Stated differently, the first law says that an object on the ground will continue to be on the ground unless a force acts on it and causes it to move. Once it is put into motion, it will continue to move—in a straight line at constant velocity—until something else exerts a force on it. If we throw a ball up into the air, for example, it will return to the ground because the Earth is exerting a gravitational force on it; if we go into deep space, however, where the force of gravity is negligible, and throw the same ball up, it will just keep going up because no forces are acting on it.

The second law has great predictive power and lies at the heart of classical physics. Momentum is, roughly speaking, what you get when you multiply mass and velocity together. If a mass stays constant, the rate of change of momentum is therefore just the product of mass and acceleration. If we know the mass of an object, we can calculate how much force it will take to get the object to accelerate at a given rate. What's more, if we know the acceleration of an object, we can predict its velocity and position at any later time.

Finally, Newton's third law states that there is no such thing as a free lunch. For every force, an action, there will be an equal and opposite force, a reaction, to balance it. Right now, the Earth's gravitational field is exerting a downward force that is holding you in your chair. At the same time, the chair is exerting an equal but opposite upward force on you to keep you from crashing through the chair and hitting the floor. As long as these two forces add up to zero, your body will remain at rest.

Newton's three laws get us halfway to the solution, but we still have to understand the nature of the gravitational force between two objects before we can solve the complete problem of getting our spacecraft into orbit around the Earth. Newton's second law related force to acceleration, but it did not define the force of gravity. Fortunately, Newton provided the answer to that part of the problem separately with his *universal law of gravitation*, or simply the *law of gravity*. The law of gravity is conceptually quite simple:

> Any two objects having mass will exert a gravitational force on each another that is directly proportional to the product of the two masses and inversely proportional to the square of the center to center distance between them.

In other words, Newton is saying that you feel a downward force on your chair because all of the mass in the Earth is attracting you. He is not discovering anything new here, he is simply describing in mathematical terms what we already knew so we could generate numerical answers. The attraction of two objects to each other is a fundamental property of nature, like mass, but it is a property that can be measured and predicted.

We can also ask if the law of gravity would predict that the ceiling above us also exerts a gravitational force on us in the upward direction. The answer is yes, it does, but the mass in the ceiling is so small in comparison to the mass of the entire Earth (about 13 million billion billion pounds) that the Earth's gravitational pull easily wins. From Newton's second law, we know that the force of gravity exerted by any planet is dependent purely on the planet's mass and our distance from it. (The relative sizes of the planets in our solar system are shown in table 1.)

Table I
Relative Size of the Planets in Our Solar System

| Planet | Average Distance from Sun | | Equatorial Diameter | | Weight | | Surface Gravity | |
	Millions of Miles	Relative to Earth	Miles	Relative to Earth	10^{24}lb	Relative to Earth	Ft/s^2	Relative to Earth
Mercury	36	0.39	3,032	0.38	0.79	0.06	12.4	0.39
Venus	67	0.68	7,519	0.95	10.67	0.81	28.2	0.88
Earth	93	1.00	7,954	1.00	13.19	1.00	32.1	1.00
Mars	142	1.52	4,219	0.53	1.46	0.11	12.2	0.38
Jupiter	483	5.19	88,860	11.17	4,192.00	318.0	75.1	2.34
Saturn	889	9.53	74,568	9.38	1,250.00	95.1	29.7	0.93
Uranus	1,783	19.13	32,189	4.05	192.00	14.5	25.5	0.79
Neptune	2,800	30.00	30,759	3.87	227.00	17.2	36.1	1.12
Pluto	3,666	39.33	1,864	0.24	0.13	0.01	1.0	0.03
Moon	—	—	2,162	0.27	0.27	0.01	5.5	0.17
Sun	—	—	864,989	108.75	108.75	332,776	899	28.0

Figure 15. Astronaut John Young leaps on the Moon on *Apollo 16*. Although the astronauts weighed over 300 pounds in their space suits on the Earth, the Moon's smaller gravitational field made them feel as though they weighed only 50 pounds. (Courtesy NASA)

According to Newton, the weight of a person will change depending on which planet he or she is on. For example, on the Earth an astronaut weighing 200 pounds and wearing a 100-pound space suit carries around 300 pounds of weight. That same person and suit on the Moon, where the force due to gravity is one-sixth that on the Earth, would weigh only 300/6 = 50 pounds. This is why the Apollo astronauts looked so light—because they were (fig. 15). This is also why Alan Shepard could hit a golf ball "miles and miles and miles" on Apollo 14. Send the same astronaut to Mars and he would weigh 125 pounds, but on Jupiter he would weigh 770 pounds. That is, if he could stand on Jupiter, he would weigh 770 pounds; but Jupiter's surface is not solid, so there is nothing to stand on.

Newton's laws of motion and the law of gravity provide two-thirds of the pieces necessary to solve the problem of getting a spacecraft into orbit, but we still need to explain why a satellite, like a ball, doesn't simply plummet back to Earth. To make things simple, suppose the orbit is circular. Without knowing anything about the nature of the force acting on a satellite, mathematicians still know that there has to be a specific relationship between the spacecraft's velocity, its altitude, and its acceleration in order for it to move in a circle. In other words, if we know the acceleration of the spacecraft, which we can calculate from the force of gravity, and we know its desired altitude, we can determine the unique *orbital velocity* that it must reach in order to stay in this orbit. If its velocity is greater than this, or lesser, the motion isn't circular but elliptical. The motion can be circular only if the motion of the spacecraft exactly counteracts the acceleration due to gravity. This is the last piece of the puzzle. Newton combined the mathematical relation for circular motion with the law of gravity and the definition of force and saw that orbital velocity for the Earth was about 18,000 mph. As shown in table 2, orbital velocity is thousands of times greater than any movement experienced on the surface of the Earth.

Newton understood that, to get something into space, you just have to throw it fast enough. This is not that big a stretch from our everyday experience. If Sammy Sosa hits a baseball, common sense tells us that the harder the ball is hit, the further it will go before it falls back to the ground. Hitting a baseball 300 feet requires a speed of about 70 mph. To clear a 400-foot fence takes 80 mph. To make it circle the globe completely—a true "shot heard 'round the world"— Sammy would have to hit it hard enough to attain a speed of 18,000 mph. This velocity is so great that a spacecraft can make a complete

Table 2
Comparison of Velocities on the Surface
of the Earth and in Orbit

Activity	Velocity (mph)
Walking	4
Running	10
Driving	55
Thrown baseball	90
Fastest train	320
Large airliner	600
Fastest airplane	2,200
Orbital velocity	18,000

orbit around the Earth, including one sunrise and one sunset, in about one and a half hours.

If we look at the problem in more detail, we see that a side effect of being in orbit is weightlessness, sometimes called *zero gravity* or simply *zero-g*. An orbiting astronaut always has mass. She is being accelerated by the Earth's gravitational field so she still has weight, but in orbit she would not be able to feel it. As she falls closer to the Earth due to the Earth's gravitational field, her orbital velocity will also move across. The net result is that she will always stay the same distance from the center of the Earth. Because she is always falling toward the Earth—a sky dive that never ends—she can't feel her weight and is "weightless." In truth, weightlessness, or zero-g, would be more appropriately termed "apparent" weightlessness, or *microgravity*. There are still small gravitational forces present, but they are about one million times smaller than what we are used to.

You can experience zero-g right now. All you have to do is stand on your chair and then jump. Between the time you leave the chair and hit the floor, you are falling under the acceleration of gravity. Since your feet are not in contact with anything, you cannot feel the force of gravity. You are weightless. How is this different from what the astronauts feel? The difference is time: it takes only a fraction of a second for your feet to hit the ground after you leave the chair. You might feel a second or so of weightlessness if you go on a roller-coaster in an amusement park, but the astronauts experience weightlessness all the time. To get used to this feeling—or to film movies such as *Apollo 13*—we can use an aircraft, climbing to a high altitude, then putting the aircraft into a parabolic dive. During the dive, the occupants of the aircraft can get a couple of minutes of zero-g. As we'll see later, when people's bodies are given time to adjust to the fact that they're weightless, it can lead to some interesting medical side effects, such as space sickness. This is why the aircraft that the astronauts use to experience zero-g has a special name: the "vomit comet."

As we saw earlier, once we get to "space" there are many different places we can choose to visit. The majority of all spacecraft are launched into Low Earth Orbit, or LEO, roughly between 50 and 500 miles altitude, usually because many missions are content just to get to "space." Once we are above the atmosphere, there is not always a clear advantage to be gained by going any higher. Of course, just as with a car, the farther you go, the more fuel you need. LEO produces the lowest gas bill and allows us to place the largest payloads, such as the International Space Station, into orbit.

But there are also some perfectly good reasons for going to higher orbits. As in mountain climbing, the higher you are, the more you can see of the Earth at any given point. But the higher the orbit, the longer it also takes the satellite to go around the Earth. As Kepler knew, increase the orbital radius by a factor of four, and the orbital period increases by a factor of eight. This means the spacecraft has more time to view any particular region of the Earth. If the spacecraft is on a remote-sensing mission, the designers must trade off the advantages of being in a lower orbit, where the spacecraft is closer to what it's trying to see but passes overhead very quickly, and being in a higher orbit, where it is farther away but has more time to look.

A desirable spot is the geosynchronous orbit, at about 22,370 miles up. Spacecraft at this altitude complete one rotation of the Earth in about twenty-four hours, the same time it takes the Earth to make one rotation. A spacecraft placed at this altitude above the equator therefore appears to remain stationary above a point on the Earth. This orbit is very popular for communications satellites because once they're there, they don't move (relative to the Earth). You can point your satellite dish at the satellite and know that the satellite will still be there days, weeks, or even months later. If you drive by your local TV station, you will see a bank of satellite dishes all pointing to geosynchronous satellites, all of them in a well-used orbit. On other planets, it would be possible to find a synchronous orbit as well—one where the spacecraft orbits at the same rate the planet rotates. However, as shown in table 3, this altitude is a function of the planet's size and its rotation rate, so it varies from planet to planet, from a mere 10,753 miles for Mars to a whopping 949,061 miles for Venus.

Although Isaac Newton knew how to get something into orbit over three hundred years ago, it was not until early in this century that scientists like Robert Goddard and Wernher von Braun tried to build machines to get us there.

The launch of the first satellite, *Sputnik I*, was made possible by a large SL-1 rocket, called a *launch vehicle*. Such a rocket is necessary to accelerate the satellite from rest on the surface of the Earth to orbital velocity at the desired altitude. Conceptually, designing a launch vehicle is fairly straightforward: simply acquire an incredibly large volume of high explosives—the rocket's fuel—and control the ignition of the fuel in much the same way your car's engine controls

Planet	Period of Rotation (days)	Synchronous Altitude (miles)
Mercury	58.7	153,858
Venus	243.0	949,061
Earth	0.997	22,274
Mars	1.03	10,753
Jupiter	0.409	54,276
Saturn	0.426	30,449
Uranus	0.451	21,709
Neptune	0.658	35,811
Pluto	6.39	18,494
Moon	27.3	54,047

Note: For Mercury, Venus, and our Moon, the synchronous altitude is so much larger than the planet itself that there is no practical synchronous orbit.

the combustion of gasoline. Direct the force of the combustion downward, and you have a rocket (plate 2); fail to contain the combustion, and you have fireworks, as in the tragic case of *Challenger* in 1986. To make the whole thing work, you need to have enough fuel to enable the rocket to burn long enough to accelerate the spacecraft to orbital velocities. But when the rocket leaves the launch tower, it is carrying a full load of fuel, so the problem is not quite as simple as accelerating only the spacecraft. Besides the spacecraft, you have to accelerate the launch vehicle and its full fuel tanks. Today, most of the launch vehicle consists simply of fuel. For example, the Space Shuttle is composed of an orbiter, two solid rocket boosters, three main engines, and a main fuel tank (fig. 16). The orbiter has an empty weight of almost a quarter of a million pounds. On top of this, there is the payload—the astronauts and their experiments, which can add another 50,000 pounds. Consequently, the Shuttle's

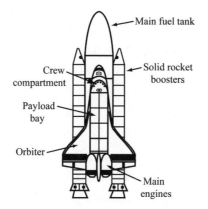

Main fuel tank

Crew compartment

Solid rocket boosters

Payload bay

Orbiter

Main engines

Figure 16. The Space Shuttle, or Space Transportation System (STS), is composed of a main fuel tank, solid rocket boosters, and orbiter. The orbiter and solid rocket boosters are reused. The main fuel tank falls into the Atlantic Ocean and sinks.

rocket motors would have to exert a thrust greater than 300,000 pounds just to lift the orbiter off the launch pad.

At the start of a launch, the three main engines on the Shuttle can exert over one million pounds of thrust. This is clearly more than enough to lift the orbiter, but let's not forget the fuel. The main engines have to burn for about eight and a half minutes to get the orbiter into "space," requiring over 500,000 gallons of liquid hydrogen and liquid oxygen. (A subtle point in rocket design is that, because of the lack of air in space, many rockets must take not only their fuel—the liquid hydrogen—but also oxygen to keep the fires burning.) The weight of the propellants is over 1.5 million pounds, but we also have to add the weight of the main fuel tank that holds it all. The orbiter, payload, fuel, and fuel tanks weigh almost 2 million pounds. This weight is far more than the thrust from the three main engines alone can handle, so two solid rocket boosters, so named because their fuel is solid rather than liquid, were added to provide additional thrust. These generate over 3 million pounds of thrust each, so both of them, when combined with the three main engines, can lift over 7 million pounds. This is more than enough to get the Shuttle off the ground and on its way.

The solid rocket boosters burn for only two minutes, but at this point the Shuttle has burned off enough fuel, and lost enough weight, so that when the boosters are ejected the three main engines alone can continue to accelerate the Shuttle the rest of the way into orbit. This concept, known as *staging*, is often used to lessen the fuel requirement for the rocket. Tossing the boosters overboard when they are no longer needed eliminates unnecessary weight. The same concept was used by the Saturn V rockets, which required three separate stages to get the Apollo spacecraft into orbit and on its way to the Moon. Unlike the spent stages of the Saturn V rockets, which crashed into the Atlantic Ocean and sank to the bottom, the Shuttle boosters parachute softly into the Atlantic and float. A recovery team tows them back to land, where they can be refueled and reused on later launches.

For the remaining six and one half minutes of the launch, the main engine's job becomes easier and easier as more fuel is spent and the combined weight of the orbiter, fuel tank, and fuel decreases. But this loss of weight can lead to another problem for the crew. As the mass of fuel decreases, if the thrust from the main engines stays constant, the acceleration felt by the crew will increase. If the acceleration is too great, it can cause loss of consciousness or even death. The Space Shuttle must throttle back its engines as the fuel is spent so that the acceleration the crew feels is not excessive. The Shuttle is designed so that the crew doesn't feel more than 3-g's during launch.

As we have seen, about 94 percent (over 4 million pounds) of the weight of the Space Shuttle is the propulsion system needed to get the other 6 percent into orbit. This is an incredibly high percentage, especially compared to your car, whose fuel would typically account for less than 15 percent of its weight. It is of critical importance to today's spacecraft designers to get this percentage down. Heavier spacecraft require more fuel, meaning a larger launch vehicle, to get into orbit. This reality stems from Newton's second law and is not really surprising. If we have to carry only a small package to a friend 100 miles away, we can fit it into a small fuel-efficient car, and the total fuel required for the trip will be minimal. If the package is too big to fit in the car, however, we have to tow a trailer, and this will require more fuel. If the package were large enough to require an eighteen-wheel tractor-trailer rig, the fuel costs would go up dramatically. It's the same way with rockets. Rocket engines deliver a constant force. Dividing this force by a larger mass (the spacecraft), results in a smaller acceleration. Less acceleration implies that the spacecraft will take longer to reach orbital velocities, which means more fuel is required. If the rocket is too small to carry enough fuel for the journey, a larger rocket must be found.

Because of the increased need for safety, manned spacecraft launches are very expensive. The total cost of a Space Shuttle mission can approach $1 billion. The development costs for the Saturn V launch vehicle, the rocket used to boost the Apollo spacecraft and Lunar Module to the Moon, were about $25 billion back in the late 1960s. In comparison, the largest rocket in the U.S. inventory for satellite launches, the Titan, is advertised to cost a mere $125 million per launch. One of the smallest launch vehicles, the Pegasus, cost only $45 million to develop and charges about $10 million per launch.

It is easy to understand why one of the golden laws, the prime directive, if you will, of spacecraft design is to "make the spacecraft as light as possible." Lighter spacecraft can be launched with smaller launch vehicles that are far cheaper. But, as Murphy's law might have predicted, anything an engineer can do to make a spacecraft lighter also makes it more expensive. Using aluminum instead of steel will save weight because aluminum is lighter, but it costs more. Using a graphite carbon epoxy material is lighter still, but also more expensive.

Roughly speaking, the larger an object, the more it weighs. This is true of automobiles, airplanes, and spacecraft, so, to build a lighter spacecraft, we need to make it smaller. The Mercury space capsules of the early 1960s provided their one-man crew with just enough room to sit down. The astronaut had no room to move around. Gemini doubled the space by adding a second chair, but also a second person to occupy it, so there still wasn't any room to move around. Apollo added a third seat, a third crew member, and just enough space to get in and out of the Lunar Module. Finally, the Space Shuttle started to add a bit of living room. However, the entire crew compartment provides only 2325 cubic feet of pressurized volume, the equivalent of a room that measures about 17 by 17 feet with an 8-foot ceiling. Try to cram eight or nine people into this space and it gets quite packed. Fortunately, zero-g helps because you are not confined to the floor—astronauts can make use of the walls and ceiling as well. The Mir Space Station provided about 13,500 cubic feet of volume, and the International Space Station will provide 43,000 cubic feet when fully assembled. It is not a lot of room for a long-duration mission, but far better than the Mercury vehicle.

Accelerating to 18,000 mph will get you into Earth's orbit, but it will not get you completely away from the Earth. To travel to the Moon and beyond, one must gain enough velocity to escape the Earth's gravitational pull entirely. This requires even greater speed than the orbital velocity we found earlier. When we calculate the *escape velocity* required, we find that it is about 25,000 mph for the Earth. Consequently, a spacecraft can escape the Earth's orbit and move on to other planets in the solar system only by continuing to accelerate past orbital velocity to escape velocity. There is a key

Table 4
Escape Velocity for the Planets of Our Solar System

	Escape Velocity	
	mph	*Relative to Earth*
Mercury	9,619	0.38
Venus	23,042	0.92
Earth	25,055	1.00
Mars	11,185	0.45
Jupiter	133,104	5.31
Saturn	79,639	3.18
Uranus	47,425	1.89
Neptune	52,794	2.11
Pluto	2,013	0.08
Moon	5,324	0.21
Sun	138,249	55.18

difference between the Space Shuttle and the Apollo Moon rockets: the Shuttle is designed to reach orbit only, and barely exceeds the 18,000-mph orbital velocity, whereas the Apollo spacecraft had to continue to accelerate to a speed of almost 25,000 mph. In fact, the *Apollo 10* Command Module set the record for the fastest speed by a man-made object at 24,791 mph. As you can imagine, this takes even more fuel than just getting into orbit. Other planets have different escape velocities, ranging from about 2000 mph for Pluto to over 133,000 mph for Jupiter (table 4).

So far, we have looked at how to get into orbit from the surface of the Earth and how to escape the Earth entirely. But if we want to go into orbit around another planet from interplanetary space, we still have a difficult challenge in front of us. A good example is the Apollo mission to the Moon. The spacecraft had to get to Earth orbit, accelerate to escape velocity for the trip to the Moon, and then be captured by the Moon's gravitational field. Now, the escape velocity for the Moon, 5,324 mph, is far less than the escape velocity for the Earth. This difference means that any spacecraft that has managed to escape the Earth's gravitational field will be traveling far too fast to be captured by the Moon's gravitational field, so somehow the spacecraft has to be slowed down. On the Apollo missions, a rough-and-ready method was used. When approaching the Moon, the spacecraft was turned around—so that the rocket motor was pointing in the direction of travel—and then the rocket motor was fired to slow down the Apollo spacecraft and allow it to enter the lunar orbit. To return to the Earth, the spacecraft had to accelerate back to lunar escape velocity for the trip home. This is the kind of problem that the Starship *Enterprise* would face routinely. Cruising at warp factor 10 might not require any energy, once you get there, but slowing down to a speed of about 18,000 mph to get into Earth orbit would take far more energy than would be required to launch the *Enterprise* from rest. As the *Enterprise* is a very massive vehicle—far larger than any spacecraft we can build today—the energy costs of orbital maneuvers would be (no pun intended) astronomical.

To reduce their fuel costs, interplanetary spacecraft can get help from a planet's gravitational field, as was done by the Voyager missions to the outer planets. The Voyager spacecraft was sent to Jupiter on a trajectory that would allow Jupiter's gravitational field to swing it in the direction of Saturn. Then Saturn's gravitational field was used to swing the spacecraft in the direction of Uranus, and so on. In many cases, the spacecraft can actually pick up speed during the fly-by of a planet. Gravitational fly-bys were used to alter the trajectories of virtually all missions to the outer planets (fig. 17). This increased kinetic energy does not come free. Because energy and momentum have to be conserved, we speed up the spacecraft by slowing down the planet. But don't worry. Speeding up a spacecraft

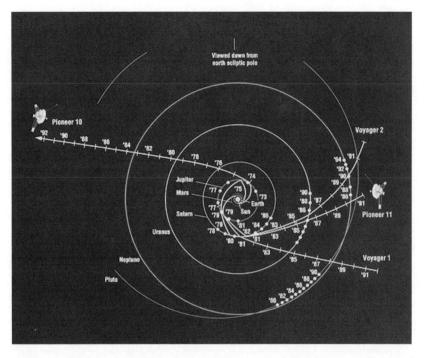

Figure 17. All of the missions to the outer planets used the planet's gravitational field to both speed the spacecraft up and alter its direction of motion. A grand tour of all the planets, as *Voyager 2* experienced after its launch in 1978, is only possible once every 180 years. (Courtesy NASA)

by 1000 mph would slow the Earth down by only about the width of one molecule per hour, so it would not be noticeable to us. The Galileo spacecraft was designed to be launched by the Space Shuttle, with a never-before-used rocket carried along to give it an extra push after being released from the Shuttle's payload bay. After the *Challenger* disaster in 1986, it was decided that the new rocket was too dangerous to carry on the Shuttle, so a less powerful rocket had to be used for the final push. The replacement rocket did not have enough power to push the spacecraft directly to Jupiter, so a series of gravitational-assist fly-bys of Venus, the Earth, and Mars were used to give the spacecraft enough speed to get to its destination.

Planetary encounters can also be used to slow down the velocity of a spacecraft. For example, NASA is currently studying the Small Solar Probe, a spacecraft designed to pass within four solar radii of the Sun's center and gather data on the environment near the Sun's surface. Getting to the Sun from the Earth is not as easy as it sounds. Even though the escape velocity of the Earth is 25,000 mph, the velocity of the Earth around the Sun is almost 67,000 mph. Suppose the spacecraft is accelerated to escape velocity away from the Earth in the direction opposite the Earth's motion around the Sun. From the Sun's perspective, the spacecraft has a velocity of $67-25 = 42,000$ mph. This is large enough to keep the spacecraft in orbit around the Sun at an altitude far higher than the four solar radii we want. To get the spacecraft to pass close to the Sun, we have to slow it down even more, so it gets sucked in by the Sun's gravitational pull. Strangely enough, it turns out that it actually takes less fuel to send the Small Solar Probe away from the Sun to Jupiter than it does to slow it down and send it straight to the Sun. For once at Jupiter, the giant planet's gravitational field can be used like an emergency handbrake to slow the spacecraft velocity so that it will then fall to within four radii of the Sun's center.

Exploring the planets with satellites is all well and good, but if we truly want to send humans across the solar system, we need to work on getting them back home. Otherwise, they will end up stranded like Major Tom in David Bowie's cult song, "A Space Oddity." Until somebody invents a transporter device capable of beaming us down instantaneously, we have to rely on the tried and true method of reentering the Earth's atmosphere. Our atmosphere exerts a drag force on a spacecraft the same way that the air exerts a drag force on a car going down the highway. To get back home, we can use this drag force to slow down the spacecraft, which will cause it to fall to a lower orbit. The atmosphere becomes denser as you get closer to the Earth's surface, so a lower orbit means more drag. More drag means more slowing, so we set up a vicious cycle. If done correctly, the spacecraft can be made to reenter on an accurate trajectory and can land on a runway, just like the Space Shuttle. Using the atmosphere, the Space Shuttle is able to slow down from an orbit velocity of 18,000 mph to a landing speed of about 225 mph, about twice as fast as an airliner. Although, as one person described it, it's not so much a landing as a controlled plummet.

In addition to slowing the spacecraft down, the atmosphere also heats it up significantly. The temperature of the outer surfaces can easily reach 2700° F. The earlier missions—Mercury, Gemini, Apollo, Soyuz—all used heat shields that would literally burn away during the most intense periods of heating. The heat shields, made of carbon, had to be thick enough to withstand only a single entry. The Space Shuttle uses silicon tiles to accomplish the same thing, the advantage being that the tiles do not burn away and can be reused. There was a significant amount of anxiety on the first Shuttle mission when some tiles fell off and left visible gaps in the outer surface (fig. 18). Fortunately, the Shuttle can withstand the absence of a few tiles here and there; missing tiles are fairly routine. They fall off because each tile has to be glued in place individually. Sometimes the glue fails or the tile is knocked loose by falling ice that has condensed on the super-cool fuel tanks during launch. Future generations of spacecraft are exploring the use of metallic thermal protection systems. Unlike the Shuttle tiles that must be glued in place, the next-generation Reusable Launch Vehicle (fig. 19), called the X-37, is designed with permanently bolted metallic plates, backed by several inches of insulation. The metallic plates may never have to be removed, so the new launch vehicle could be readied for launch much more quickly and cheaply than the Space Shuttle.

Figure 18. The Space Shuttle lost some tiles during its first launch (the dark spots on the left engine pod). Fortunately, the Shuttle can lose a few tiles and still return safely. (Courtesy NASA)

Reaching the surface of a planet with a thick atmosphere, such as Earth or Venus, causes tremendous heating by the atmosphere, so heat shields are mandatory for such landings. Coming down on a body that has no atmosphere, like the Moon, creates different problems. To slow the spacecraft down, it has to carry additional fuel to provide the thrust necessary to touch down safely on the surface. It is a Catch-22 situation: when entering a planet that has an atmosphere, one can use the drag force to slow the spacecraft without burning fuel, but at the expense of reentry heating; entering a planet with no atmosphere means no heating, but the entry requires more fuel.

Figure 19. NASA's X-37 is an experimental prototype for the next-generation reusable launch vehicle (RLV). It will be designed to be launched more quickly, and less costly, then the Space Shuttle. (Courtesy NASA)

At the moment, future space travel is up against cash-flow problems. The cost of today's launch vehicles is enormous, so spacecraft are accelerated to orbital velocity (or escape velocity, as the case may be), but no more. Besides reducing travel time, there are few other benefits to traveling faster. Unfortunately, as we start to cruise our cosmic neighborhood, we find that the distance between planets is so vast that it would take years just to get to the edge of our solar system. Clearly, a faster way to go must be found before we can explore the galaxy in person.

Several other possibilities have been examined to accelerate spacecraft, such as using light from the Sun. Because light contains energy, it also contains momentum. Consequently, light from the Sun exerts a tiny force on a mirror when it is reflected, but it is so small that, in the presence of a gravitational field like the one on the Earth, the light force is overwhelmed by gravity and is unnoticeable. But in the weightlessness of space, it would be large enough to make things move. Although they are theoretically possible, these light sails would have to be several miles in length, to capture light for a large enough force; it would also have to be only a few atoms thick to keep its weight at a reasonable level to make it a practical tool. Another method would be to boost the light sail by using laser light from the Earth. Laser-powered spacecraft have already been tested on Earth, inside our atmosphere, but remain to be tested in space. We still need another means of propulsion before we can make the journey to the planets in days or weeks instead of months or years. To understand the realities of high-speed interplanetary travel, we need to understand something about Einstein's theory of relativity.

Watch virtually any episode of *Star Trek*, *Star Wars*, or other futuristic space adventures, and you will be bombarded with references to starships traveling faster than the speed of light. Such ultra-high-speed travel is more than just a convenience when exploring the stars—it is a necessity due to the vast distances involved. Even though a spacecraft may leave the Earth's gravitational field at a speed of 25,000 mph, it will take months or years just to travel to other planets within our solar system at that speed (table 5). Even at its current speed of 38,000 mph, the *Pioneer 10* spacecraft—one of the man-made objects farthest from the Earth—would take over 2 million years to reach the nearest star along its trajectory to the constellation Taurus. Even if we headed directly toward the star closest to our Sun at escape velocity, our spacecraft would decompose into nothingness long before it completed its 115,000-year journey. Because of our current limitations, travel to other star sys-

Table 5
Travel Time to the Planets of Our Solar System

Planet	Closest Distance from Earth (million miles)	Travel Time	
		At 25,000 mph	At Speed of Light
Mercury	57	95 days	5.1 min
Venus	26	43 days	2.3 min
Earth	—	—	—
Mars	48	80 days	4.3 min
Jupiter	390	1.78 years	34.9 min
Saturn	795	3.62 years	1.20 hours
Uranus	1,690	7.70 years	2.52 hours
Neptune	2,703	12.31 years	4.16 hours
Pluto	3,573	16.27 years	5.32 hours
Moon	0.237	9.36 hours	1.27 sec
Sun	93	155 days	8.3 min

tems is simply not feasible. The only possible way to speed up the trip is to speed up the spaceship.

The closest distance from Earth to Mars is about 48 million miles. Even at the escape velocity of 25,000 mph, a one-way trip would require about three months' travel time. To make matters more complicated, the Earth and Mars are continuously in motion around the Sun. If a quarterback wants to throw a wide receiver a pass, he has to aim not for where the receiver is, but for where the receiver will be when the ball arrives (fig. 20). It is the same in space travel. Therefore, the actual path a spacecraft from Earth has to follow to get to Mars is much farther than the straight-line distance, and the usual travel time is about nine months (fig. 21)—just long enough for an astronaut who is pregnant at takeoff to give birth to the first Martian. If the spacecraft could accelerate at 1-g to the halfway point, then turn around and decelerate the rest of the way, the three-month trip would be reduced to about a day and a half. This time could be

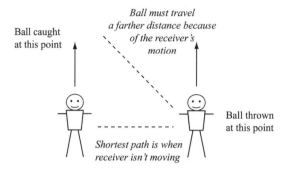

Ball caught
at this point

Ball must travel
a farther distance because
of the receiver's
motion

Ball thrown
at this point

Shortest path is when
receiver isn't moving

Figure 20. If you want to pass me a football while I'm running down the field, you have to aim not for where I am, but for where I will be when the ball reaches me.

quite acceptable to future space travelers: this is about the same time it would take now to get from New York City to to Los Angeles by train nonstop. This method would also have the added benefit of providing the crew with an Earthlike 1-g environment to live in, rather than weightlessness, so travelers would not have to worry so much about possible space sickness. But, alas, all this is beyond today's technology. Right now, our trip at 1-g would necessitate a fuel tank about the size of the Earth just to make the one-way journey to Mars, and the stars are even farther away.

Let us take a look at the challenge we face in traveling to the closest star to the Sun, Proxima Centauri. (It is visible only in the Southern Hemisphere, so don't try to find it in the night sky if you are very far north of the equator.) Because of the great expanse of space, the distance between stars is often measured in *light years*, where a light year equals the distance that light travels in one year. The speed of light in the vacuum of space is an impressive 186,000 miles per second, almost 670 million miles per hour. The Sun, which is roughly 93 million miles away, is a distance of 8 light minutes from us. Compare that to a travel time of 163 years, (plus pit stops) at a typical freeway speed of 65 mph, and you get a better

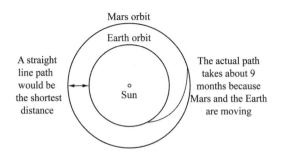

Figure 21. A trip to Mars requires the spacecraft to travel a path that is far longer than the straight-line distance.

feeling for just how much faster the speed of light is than our usual travel speeds.

In one year, light travels about 6 million million miles. Proxima Centauri, which is 24 million million miles away, is therefore about 4 light years from Earth. How do we know this? By measuring its parallax, which we discussed in the first chapter. Unfortunately, the parallax measurements of even the closest stars are tiny. The parallax of Proxima Centauri is only about 0.75 arcseconds, or about 0.0002 degrees—the apparent size of a quarter viewed from a distance of 4 miles. Although the term "parsec" was used in the movie *Star Wars* as a unit of time, parsec is used in stellar travel as the distance from which a star has a parallax of one arcsecond using the Earth's orbital diameter as a baseline. A parsec is about 3.26 light years. Only the stars closest to Earth have a parallax that is large enough to be determined even with the most sensitive of instruments. A variety of indirect techniques have to be used to infer the distance to the other objects in the Universe.

Because of Proxima Centauri's distance, we know that it takes its light four years to reach us here on Earth. This means we're seeing Proxima Centauri not as it is today, but as it was four years ago. An interesting side effect of the finite speed of light is that astronomers can look back in time as they look at objects that are farther and farther away. A star cluster that is 10,000 light years away appears to us as it was 10,000 years ago; a similar star cluster that is 100,000 light years away looks the way it was 100,000 years ago; and so on. This makes it possible for astronomers to compare how star clusters might have evolved over the ensuing 90,000 years.

If we managed to travel to Proxima Centauri successfully, it would take four years for a radio signal carrying the news of that accomplishment to reach our friends on Earth. This distance would create quite a communications problem. At the speed of light, a message would take one second to reach the Moon. Though this delay is noticeable, it is not enough to cause a big inconvenience. In the early days of the telephone, there was always a few seconds' delay between the time you spoke and the time the message was received at the other end. A delay of a second or two is tolerable. Mars is about 20 light minutes away. When Pathfinder landed on Mars in 1997, it took about twenty minutes for news of that event to reach us here on Earth, and another twenty minutes for Pathfinder to hear our reply. A forty-minute delay in conversation would put a great damper on real-time communications. This delay also applies to relaying important instructions, which might mean the difference between a successful and disastrous mission.

As we saw earlier, at the escape velocity of 25,000 mph, the trip to Proxima Centauri, which is four light years away, would take over 115,000 years. If we were able to accelerate at 1-g for half the trip, then decelerate for the other half, the one-way trip time would take about four years. (It is just coincidental that our 1-g rocket ship makes the four-light-year journey in four real years. This is not a rule of thumb for space travel; the numbers just happen to match up in this particular case.) With the return trip, the crew would be away for at least eight years—plus whatever time they spent in orbit around the star. It's doubtful such a trip would attract many volunteers, but an eight-year mission might be feasible some day. Unfortunately, a trip of this type would not be possible now even if we had a far more efficient fuel than we can manufacture today and even if we could build a planet-sized fuel tank to hold it. The reason is relativity.

After the spacecraft accelerated outward for two years at 1-g, its velocity would be about 380,000 miles per second, or twice the speed of light. This would have been fine with Isaac Newton, but not with Albert Einstein. Einstein is probably the most famous physicist of all time. He was even named *Time* magazine's "man of the century" in 1999. In 1905, while working as a clerk in a patent office in Switzerland, he introduced the special theory of relativity, based on two assumptions that sound very simple. Suppose you have two physicists who are moving in straight lines at constant speed. Einstein said, first, that the laws of physics must be the same for both physicists, which makes perfect sense. His second assumption is more daring: that the speed of light, as measured by both physicists, must also be the same. This second point was far more difficult for the physics community to swallow.

The first assumption states that any experiment performed by two different people would give the same result even if one of them were moving—with constant velocity—relative to the other. In other words, it shouldn't matter if you are standing still and I am flying overhead in a straight line; both of us should get the same result for any experiment we do. This is easy to verify. Whether you drop a book while standing on the surface of the Earth or while flying overhead in an airplane at a speed of 600 mph, gravity will accelerate it downward at the same rate.

Einstein's second assumption states that both of us would measure the same velocity for the speed of light, even though you are standing still and I am moving. This seems bizarre. If you are standing still and I pass you on the freeway at 60 mph, while at the same time a second car passes you with a speed of 120 mph, I should expect to see the second car moving at 60 mph relative to me, right? Einstein is saying that if the second car is moving at the speed of light, then I would also perceive it as moving at the same speed of light—my 60 mph velocity wouldn't matter. Although this seems intuitively wrong, Einstein is right! The reason he's right is that the very nature of the way in which we perceive the Universe depends on the light that we receive from the objects around us. How do you determine the velocity of a car that rushes past you on the freeway in the first place? You simply note the time when the car appears to be at a certain point on the road, then you note the later time when it appears to be at a different point. The ratio of distance between the points to the time elapsed is, by definition, the car's velocity. This method works fine until the object in question has a velocity that is near that of the speed of light—then it breaks down because our perception of time and distance changes.

Suppose a mirror stands a fixed distance away from us and we want to know exactly how far away it is. However, instead of a measuring stick all we have is a firecracker and a match. How can we use the firecracker to tell us how far away the mirror is? We can measure the time it would take for the light from the firecracker to travel from us to the mirror and back. This is simply the mirror's distance divided by the speed of light. In other words, we can measure the distance to the mirror by monitoring the time it takes for the ray of light to reflect off the mirror and return to us. Unfortunately, this simple scenario gets more complicated if we are moving.

Let's return to our earlier example of a quarterback throwing a ball to a wide receiver. This may be easier to visualize if we think of two people playing catch rather than shining beams of light on a mirror. If you and I are standing still and tossing a football back and forth at some fixed speed, the time it takes for the football to leave your hands and reach mine depends only on the distance between us. If both of us start jogging down the street at the same speed so that the straight-line distance between us doesn't change, in order for you to pass me the football you have to throw it not to where I am, but to where I will be when the ball gets there. It is obvious that the path the football must travel is longer if we are moving than if we are standing still. Since the path is longer, it must take a longer time for the football to pass back and forth if the speed of the football is fixed. If we replace the football in our analogy with a light beam, which travels at a fixed speed, we must conclude that it will take light longer to travel to the mirror and back if the mirror and light are moving than if they are standing still. In other words, when we are moving, our perception is that it takes longer for events to occur. This is called *time dilation*. A consequence of relativity is that a moving observer will see time pass more slowly than an observer at rest.

Two things are important to bear in mind. First, geometry is not important. It does not matter if we're moving to the right, the left, up, down, in or out—the result is the same. Second, time dilation has been rigorously verified by many experiments: we have proof that it is real. In this day and age, atomic clocks can be constructed that are accurate to one-billionth of a second. If we take one clock on a nonstop airplane trip around the world and leave another at home, when we return we'll find that the clock taken on the trip has lost time relative to the clock that stayed put. In other words, the moving clock ran more slowly. This is not something we are used to seeing in our everyday lives because the effect is so small even for aircraft traveling at 600 mph. For orbital velocities of roughly 18,000 mph, the time difference due to dilation is only one second every three hundred years. You would have to travel at 558,000 mph for the time difference to be one second over the period of a year. This is a far greater speed than humans can yet travel, so we can't perceive time dilation with our bodies even though we can measure it in the laboratory. Nevertheless, it is real enough so that the first of a pair of twin astronauts to make it into space would be able to claim the title "the younger" until his or her twin gets a turn to fly. (Interestingly, NASA does have twin astronauts in its program, so we may actually be able to do this experiment in a few years.)

Not only does time slow down as we move faster and faster, but objects get smaller as well. This is called *length contraction* and is another consequence of special relativity. If your car is at rest and you see a truck move past you at a certain velocity, you can determine the truck's length by multiplying the velocity of the truck by the time it takes the truck to pass you, as measured on your clock. From the truck driver's perspective, you moved past him with precisely the same velocity, only in the opposite direction. But since he is moving, his clock will run slow. That means that it actually takes less time for his truck to pass you, and you will think his truck is shorter than it actually is. This is length contraction.

As with time dilation, length contraction is far too small to be noticeable at everyday velocities. For a spacecraft at orbital velocities, the contraction is about one part in 100 million. In other words, length contraction shortens the Space Shuttle Orbiter from its nominal length of about 120 feet by about 0.000,01 inches. This is far smaller than "normal" changes in length that occur because of thermal expansion and contraction. Today's space travelers usually don't have to worry about relativistic effects, but tomorrow's star travelers would. If we could continue to accelerate to a velocity that is 86.6 percent the speed of light, time will have slowed, and length will have contracted, by a factor of two.

In addition to time dilation and length contraction, relativity also shows that the speed of light is the ultimate speed limit. No object can go any faster. If you are traveling at the speed of light, you can't even look at yourself in a mirror. If you hold a mirror in front of yourself, then the mirror is also moving away from your face at the speed of light. How would the light that bounces off your face ever make it to the mirror? As Yahoo Serious said in the movie *Young Einstein* (1989), if you move away from a clock at the speed of light you would never see another second tick by because the light from the clock would never catch you—time would stand still. This is time dilation in action. As we move faster and faster, time slows down. If we could ever reach the speed of light, time would literally stop. We'd be forever frozen at that instant in time because it would take forever for the next second of the clock to tick by. Likewise, length contraction predicts that our size grows smaller and smaller as we approach the speed of light. Not only would time stop as we reach it, prohibiting us from ever seeing the next second, we would shrink out of existence as well.

As if these two factors are not enough of a problem, a third conse-quence of relativity is that as an object speeds up, its mass actually increases—by the same factor that length contracts and time dilates. In other words, by the time our spacecraft had accelerated to 86.6 percent the speed of light, its mass would have doubled. This would reduce the effective acceleration from its engines by a factor of two. Even if we have infinitely capable rockets and can double our thrust—so that the acceleration remains constant—the fuel require-ment to feed the rocket would also double, just as the spacecraft mass doubles. As we continued to accelerate the spacecraft, its mass would continue to increase, causing the fuel requirement to increase even more. Unfortunately, as our velocity increases to the speed of light, the spacecraft's mass increases to infinity. It would take an infinite amount of fuel to push our infinitely massive spacecraft, and an infinite supply of anything just isn't available. Consequently, we are doomed to travel through the Universe at velocities slower than the speed of light.

Let's come back to the problem of fuel for a moment and get a better understanding of just what might be possible one day. Modern fuels release the energy that is stored in the chemical bonds of material through chemical reactions that change the nature of the material. Gasoline, when mixed with air and compressed in your engine, releases an enormous amount of energy while a chemical reaction, called *combustion*, takes place and turns the gas-and-air mixture into the foul-smelling matter that comes out of your exhaust pipe. To get more energy out of the reaction, you simply need to release more energy out of the chemicals you put in. Quite obviously, the best you can do is to extract all of the energy that the chemical possesses to begin with. In other words, the most efficient reaction possible is one where the fuel is converted entirely into energy.

As a fourth consequence of relativity, Einstein saw that there was a definite relation between mass and energy. In his famous equation, $E = mc^2$, he showed that energy is equal to mass times the speed of light squared. In other words, mass *can* be completely converted into energy, (and vice versa). Converting mass completely into energy can release far more energy than is possible with typical chemical reactions. For example, converting a single gram of mass into energy would be enough to power a 100-watt lightbulb for over 100,000 years. Reactions that convert mass into energy are called *nuclear reactions* and take place in nuclear reactors. Nuclear warheads operate on the same principle, with the important difference being that nuclear reactors control and contain the energy-conversion process while nuclear warheads do not. Although nuclear reactors can have serious problems and can "melt down" like Chernobyl or Three Mile Island, they can't blow up like atomic bombs.

Nuclear reactors come in two flavors, fission or fusion. *Fission* reactors split larger molecules, like uranium or plutonium, into smaller molecules. When finished, the whole is greater than the sum of its parts. The mass difference between the small molecules you finished with and the large molecule you started with appears as energy. The opposite process, called *fusion*, is used to convert light molecules, like hydrogen, into heavier molecules. The mass difference between the "before" and "after" molecules appears as energy. Fusion is the process that fuels the Sun. However, in both fission and fusion reactions there is always mass left over. The ideal situation would convert all of the mass into energy, not just part of it.

This is the principle behind the *Enterprise*'s fabled matter/anti-matter engines, fueled by dilithium crystals. We're all familiar with matter: it's simply the stuff that makes up the world. The laws of physics predict that the exact opposite of normal matter, called *anti-matter*, should exist in nature. The problem with finding it is that if a piece of matter happens to come in contact with a piece of antimatter, both will be annihilated into pure energy. It's difficult to do, but scientists can create antimatter in the laboratory. If we could find a cost-effective means to do so, it would provide us with the most compact means possible of storing fuel for future journeys.

As a rule of thumb, the amount of matter/antimatter fuel needed to accelerate a spaceship depends directly on the mass of the space-craft. It also depends on how fast you want to go, relative to the speed of light. To accelerate the spacecraft up to half the speed of light would require a mass of fuel greater than the mass of the space-craft. To decelerate it back to rest would require the same amount of fuel. In other words, we're still in the same boat we're in with the Space Shuttle today. Most of the mass of star-traveling spaceships would have to be fuel. These massive interstellar spaceships would be confined to travel through the galaxy at speeds less than the speed of light. Although time dilation may make the crew feel that the journey is only a few days long, observers on Earth would grow old and die before our star voyagers could travel to all but the nearest star systems.

All of the preceding discussion stems from Einstein's special theory of relativity. As difficult as the special theory is to understand fully, it is still far less complex than Einstein's general theory of relativity. Before Einstein, we spoke of "space" and "time" separately. With special relativity, Einstein showed that we should really speak of "spacetime" as one single concept. That is, our concepts of space and time are so closely related that a single set of mathematical relations can be used to describe them. Einstein set about trying to derive the mathematical relations that would describe how spacetime was affected by the presence of mass: he wanted to develop a more general set of equations that would, among other things, include the effects of gravity. Einstein succeeded in laying the groundwork that describes how mass, due to a star, a planet, or even to you and me, alters space and time. Out of this work came some predictions that were as equally startling as time dilation and length contraction.

First, the general theory explained a slight wobble, or precession, in the orbit of Mercury. Kepler predicted that the orbit should be a perfect ellipse, but observations showed a slight deviation that no one, until Einstein, was able to explain. This was a sure sign that Einstein was on the right track. What was hard to believe was that the general theory also showed that mass bends light. This was almost inconceivable.

Light has no mass, so, according to Newton's law of gravity, light should not feel a force from objects having mass and thus could clearly not be bent by them. But again, Einstein was right. He showed that mass literally bends spacetime, and light, having no alternative but to follow the shortest distance between two points, must respond when that straight-line path gets bent. In 1919, astronomers tested this theory by watching the stars very near the edge of the Sun during a solar eclipse. Sure enough, the stars "appeared" to be displaced from where they ought to be—a sign that the mass of the Sun had indeed bent the light from the stars as it passed by. The Hubble Space Telescope has also found convincing evidence of light being bent by mass in its observations of distant galaxies (plate 3).

Einstein's new equations showed that the more massive the star, the more it would bend light that tried to pass nearby. People began to recall a question asked over 120 years earlier. In 1796, the Marquis de Laplace wondered if there could be an object so big that not even light could escape its gravitational field. The idea died rapidly at the time because the physics of the late 1790s had no way of even defining the problem realistically. With Einstein's theory, people now had the tools to solve it. Karl Schwarzschild showed that if a star contracts there will come a point where nothing, not even light, can escape its gravitational field. Because no light would be emanating from these objects, they were rapidly dubbed *black holes*. As hard as it may be to imagine, we have very convincing proof, due to the work of modern-day physicists like Stephen Hawking, that black holes do exist in certain parts of our galaxy (plate 4).

If we accept the presence of black holes, a place where things enter but never exit, could we also have the exact opposite—a place from which things exit, but never enter? Some people believe that Einstein's equations can be used to predict both possibilities. Logically, they call the other possibility *white holes*. If a black hole is a one-way entrance and a white hole is a one-way exit, could they be connected? Again, some people believe this might be possible, and they call the connections *wormholes*. Might it be possible one day to enter a black hole, travel across the galaxy in a wormhole, and emerge thousands of light years away in a white hole all in the span of a few minutes' time? It's a fun problem to consider, but the only way one could see if it would work is to jump into a black hole and see what happens. Unfortunately, the gravitational forces near a black hole would be so intense—by definition great enough that not even light could escape—that they would literally rip you to pieces. Even those willing to risk it would still have to make the journey to the nearest black hole using our limited technology. This means it would take 100,000 years or so if we left right now. Alas, we are doomed to explore the galaxy at subwarp speeds.

A Space Odyssey:
Spaceships and Starships

I cannae do it. Captain!
I've got to have more power!

As chief engineer of the *Enterprise*, Mr. Scott's responsibility is to maintain the ship in tip-top condition at all times. This is no trivial task, for a spacecraft has to be its own self-contained Universe. It must not only generate its own power for the duration of the mission, but it must also maintain the proper environmental controls for the subsystems and crew; it must orient and maneuver the vehicle, and so forth. All of a spacecraft's needs during its mission must be planned for in advance and built in from the start.

Although the terminology changes over time, there are several general functions—called *subsystems*—that engineers consider when designing or operating a spacecraft. Once launched, the first task for a spacecraft is to establish contact with the ground and verify that all is well. If the spacecraft cannot even manage to point its antenna at the Earth, the mission may be doomed from the start. For this reason, a critical element in spacecraft design is *attitude determination and control.*

"Attitude" is the term engineers use for the orientation of the vehicle. In space it is hard to know which way is up. Determining the spacecraft's attitude, and controlling the attitude so that the spacecraft is pointing in the right direction, is a far more difficult problem than it is on the surface, or in the atmosphere, of a planet. Regardless of whether we are in orbit or cruising in interplanetary space, simply being in space means that the spacecraft is in weightlessness. If you close your eyes right now, your senses will be able to tell you which way is down because you'll be able to sense the force of gravity acting on your body. However, if you come down with a bad case of vertigo so that you can't tell which way is up, little things like standing up or walking in a straight line become far more difficult. Spacecraft have to deal with vertigo all the time.

Just how critical the attitude determination and control subsystem is was seen vividly on the *Gemini 8* mission in March 1966. Astronauts Neil Armstrong and David Scott got to experience the near disastrous side effects of an attitude-determination and control system gone bad. The Gemini astronauts had just completed the first ever in-orbit docking with an unmanned Agena rocket and were preparing for a spacewalk when their spacecraft began to roll to the left for no apparent reason. The astronauts turned off the Agena's attitude determination and control system, hoping that if the rockets system were bad they could stop the problem by simply turning it off. However, the problem became worse. The astronauts decided to undock from the Agena before the two spacecrafts were damaged. Unfortunately, this only made the twisting and turning worse, and soon the Gemini spacecraft was "tumbling end over end" at the rate of once per second. Spinning at this rate is extremely dangerous because it can cause one to become disoriented and pass out, so the astronauts had to act quickly. They realized that one of their attitude-determination and control thrusters must be firing continuously, so they shut down the entire system and activated a separate reentry control system that allowed them to regain control of the vehicle.

Mission rules dictated that if the reentry control system is activated, the spacecraft must reenter immediately, so the mission ended abruptly, but safely, a mere ten hours after launch. Until the time of *Apollo 13*, this near disaster was the closest call for the American space program.

Much like people, spacecraft are susceptible to the small disorienting forces that we can ignore in our everyday life. Because there is no force of gravity to anchor the spacecraft to the surface of the Earth, even the smallest force can accelerate a spacecraft and cause it to move. A crew member who propels himself off the wall of the Space Station exerts a force on it (a reaction to the crew member's action) that will accelerate the Space Station and cause it to move. This motion could cause the spacecraft to start spinning or drift out of its intended orbit. Even tiny forces due to drag from the Earth's atmosphere, the Earth's magnetic field, or even the pressure of light from the Sun on the spacecraft could be significant.

Back in the 1970s, there was a cult science fiction television program called *Lost in Space*. Engineers recognize this possibility of getting lost all too well. There is little point to spending millions of dollars to get a spacecraft into orbit if the spacecraft can't tell where it is and, more importantly, where it is supposed to be once it gets there. Luckily, physics can come to the rescue. Based on the equations of motion, if you know where something is now and you know the strength of the force and how long it will last, you can predict where an object will be at any time in the future. So, if a spacecraft knows where it is, and if it is stocked up with sensors and "inertial measurement units" that monitor the duration and extent of any acceleration, it will be able to calculate where it will be in the future. Spacecraft that orbit the Earth have "Earth horizon sensors" to tell them where they are relative to the Earth. For good measure, many spacecraft also have Sun sensors and star sensors. The data from all three types of sensors are fed into the spacecraft's computer. As long as the computer doesn't crash, all should go well, because knowledge of the relative positions of the Earth, the Sun, and the stars is all that is required for the spacecraft to be able to calculate its position and orientation.

Once the spacecraft has been given some means of determining its attitude, it must also be given some means of controlling it. We're in luck if our spacecraft is long and thin, rather like a pencil, for then gravity works in our favor. It will cause the spacecraft to align itself so that it points toward the center of the gravitational field, which in this case is the center of the Earth. This is called *gravity-gradient control*. Of course, not all satellites are pencil shaped, so we need a few other choices. A football quarterback throwing a pass always tries to put a spin on the ball because the spin makes the ball travel smoothly toward the intended receiver. A baseball pitcher tends to do the same thing, because a spinning baseball can be more easily controlled. A knuckleball pitcher gives the ball no spin at all and the pitch moves all over the place. Engineers have learned from these techniques and make some satellites spin so that they continue to point in the same direction. This technique is called *spin stabilization*.

More often than not, the spacecrafts of today are three-axis stabilized, and they are likely to remain so in the future. These craft are able to maintain orientation and stability independent of any reliance on gravitational fields or spinning. This is more expensive and complex to achieve but is often required by the mission. To maintain three-axis control, designers may pick from a wide assortment of stabilizing devices such as momentum wheels, reaction wheels, control moment gyros, and magnetic torquers. Often, though, the engineer will simply rely on thrusters—small rocket motors that can be fired to exert small forces that either move or rotate the vehicle. For

Figure 22. The Space Shuttle seen docked with the Mir Space Station. The Shuttle must align itself to within one inch of the Mir, and must only be traveling a few inches per second, in order to dock safely. (Courtesy NASA)

example, the Space Shuttle reaction control system is composed of dozens of separate thrusters that are used to maneuver and orient the vehicle. When docking with the Mir Space Station, the Space Shuttle had to align its docking device within about one inch of the Mir docking ring (fig. 22). At the same time, the closing velocity between the Mir and the Shuttle had to be no more than a few inches per second, or both the Mir and the Shuttle could have been damaged. The alignment and approach were controlled with the Shuttle's thrusters.

In the future, the Starship *Enterprise* of *Star Trek* will have a very formidable problem with attitude control because of its immense size. The *Enterprise* is almost a half mile in length. Rotating a space-craft of this size and mass is far from trivial. If we wanted to do a cosmic U-turn and a 180° about-face within one minute, the outer hull of the *Enterprise* would have to move at a speed of about 40 mph. As with the forces one feels on a roller coaster, the crew members near the outer hull would be accelerated quickly and then decelerated back to rest just as quickly when the rotation stopped. In a real emergency, such as an attack by Klingons or the Borg, the crew would be flung against the walls of the vehicle with great force. Simply moving the *Enterprise* would exert a great deal of stress and strain on the structure and the crew inside it. To acknowledge this fact, the show's developers include fictional "inertial dampers" to mysteriously stabilize the vehicle against sudden forces. In reality, a spacecraft as large as the *Enterprise* would have to take great care not to rip itself apart during simple maneuvers.

If you have ever typed away on your computer and told yourself you would save your work when you finish the next paragraph, but then the power went out and wiped out a whole afternoon's work, you know how important it is to have power available when you need it. Power is everything in space travel. Without it, you can't fire spacecraft thrusters, you can't establish radio contact with Earth, and you can't run your life support systems. Without power you die. Consequently, producing, storing, and distributing electrical power onboard the spacecraft is a vital necessity.

A prime example is the near disaster on *Apollo 13*. About fifty-six hours into the mission, astronaut Jack Swigert was asked to stir the oxygen tanks that provided the crew with breathing oxygen as well as being a source of fuel for the "fuel cells" that produced power for the vehicle. A few seconds later, an electrical arc jumped across two wires and started a fire in the oxygen tank. The tank exploded and blew off the spacecraft's outer panel. The explosion shut off the oxygen supply to two of the spacecraft's three fuel cells, which immediately stopped working. The explosion also damaged other tanks and equipment, and the final fuel cell started to fail as well. If the third cell died, the command module would be left with no electricity. The only additional source of power on the command module was a reentry battery, designed for use only during the last few minutes of the mission when the spacecraft was heading for its splashdown. The astronauts rapidly disconnected the battery to prevent it from being drained. With only minutes of power—and minutes of oxygen—left in the command module the astronauts were in a critical situation. It would take them at least three and a half days to get back to Earth.

Fortunately, they had a separate spacecraft in tow. The Lunar Module was designed to operate independently of the command module, so it carried its own oxygen and power supplies. The astronauts were able to use the Lunar Module as a lifeboat for the next three days until they completed their loop around the Moon and returned to Earth. Unlike the Command Module, which was designed to support three men for a week, the Lunar Module was designed to support only two men for about two days. This meant it produced less power, carried less oxygen, and so on. The astronauts had to reduce their life support systems to a bare minimum to make sure that their meager resources would last for the return trip. With minimal power, there was no way they could afford to heat the spacecraft, and the temperature plunged to around 50° F. With minimal oxygen—which is not only the source of power and air but also, mixed with hydrogen, of drinking water—they could afford to drink only 6 ounces of water per day instead of the preferred "minimum" of 30 ounces. They also had to find a way to allow the Lunar Module's environmental control system to remove the carbon dioxide exhaled by three people instead of the two for which it was designed. Fortunately, the Lunar Module, *Aquarius*, gave them a safe ride back to Earth. They reconnected the reentry battery on the Command Module, the only part of the spacecraft with a heat shield, and splashed down in the Pacific with virtually no power—and no oxygen—remaining. Had the Lunar Module not been capable of generating power independently of the Command Module, the astronauts would have perished.

Strictly speaking, power-production techniques do not really produce power. That is, we can't create power where there was none to begin with. What we can do, however, is to convert energy from one form to another. To a scientist, power is the output of energy per unit time. A lightbulb in your home, for example, may be 40, 60, 100 watts, or higher. Its wattage is a measure of the amount of light energy it emits per second. The bulb's light energy is produced by a glowing, hot filament in the bulb. The thermal energy necessary to heat the filament comes from the electrical energy flowing through circuitry, which in turn came from a power plant that may have created the electricity from coal, gas, or even nuclear power. Spacecraft typically do not have room for large power plants, but a variety of techniques exist for the production of power in space.

To decide which technique is ideal, spacecraft engineers first have to ask some simple questions: How much power is required at the start of the mission? Will more, or less, power be needed as time goes by? Will there be peaks or valleys in the power requirement, or is it constant? Once these questions have been answered, the engineers work out just how much power is needed, and how often. Power loads are rarely constant. It takes more power to repel an attack by the Borg or chase after a Romulan vessel that just invaded the neutral zone than it does to conduct routine star-mapping operations. This variation in power consumption takes place in your house all the time. At night when everyone is asleep, there is minimal power usage. In fact, the only power consumption at night might be from the electrical alarm clock by your bed. But when you wake up and start moving about, power consumption increases. You turn on the lights, you take a shower—which means that the hot water supply is diminished and power must be used to heat more— you cook breakfast. Each event requires more power. Similarly, spacecraft power consumption is often cyclical. Communications spacecraft can be expected to carry a lot of phone calls during the day or early evening when people are awake. At night, people don't make as many calls because most are asleep. Once the power requirement is known, the optimal method of generating power can be determined.

Batteries have been, and will continue to be, a basic means of power generation for long-lived spacecraft. A spacecraft battery is not too different in concept from the battery in your car. Both types convert chemical potential energy, stored in the battery, into electrical energy. The electrical energy is used to power electrical components. There is a problem, however, because batteries provide only limited amounts of power, as anyone who has left car lights on knows. This means they are best suited for short-term use. If the mission is short, such as a ten-minute journey by a launch vehicle, a battery may be all that is needed. But if the mission must last longer than a few days, the battery that would be required would often be too large to fit inside the spacecraft. Fortunately, many batteries, such as nickel-cadmium or "ni-cad," are rechargeable. At home, you plug the battery recharger into a separate source of power—the electrical outlet in the wall—in order to recharge the batteries. To recharge batteries in space, we often rely on solar arrays.

Solar arrays, or collectors, are a popular method of generating electrical power (fig. 23). These arrays can tap into a seemingly endless supply of energy, the Sun, which radiates energy into space in

Figure 23. A solar array experiment seen in the Space Shuttle payload bay. Solar arrays are used to convert light from the Sun into electricity for the spacecraft. (Courtesy NASA)

the form of light at the rate of about 390 million billion billion watts. Only a tiny fraction of this light reaches the Earth, but that "tiny" fraction amounts to about 170 million billion watts, enough power to turn more than one billion tons of ice into steam every second. Even though a solar array may capture only a fraction of this energy, there is still a lot to go around.

Solar cells convert light energy from the Sun into electrical energy. They are constructed out of materials called *semiconductors*, such as silicon or gallium arsenide. The solar cell absorbs light from the Sun and converts the light energy into kinetic energy for electrons in the semiconductor, and electrons in motion become "electrical current." In this way, the light from the Sun is used to generate a small amount of current flow, which can then be used to power electrical equipment. The only problem is that this method is not particularly efficient, ranging somewhere between 15 and 20 percent, depending on the technology used. To generate 100 watts of electrical power, a spacecraft would need to have a solar array large enough to collect over 500 watts of light power.

Besides their low efficiency, solar arrays have a problem that is fairly obvious: they cannot work at night. To overcome this obstacle, we often use the winning combination of batteries and solar arrays. Batteries are used to power the spacecraft during nighttime, when the Sun is behind the Earth and solar arrays are useless. Luckily, nighttime in Earth orbit is only about 30 minutes long at space station altitudes, so batteries can provide a fair amount of power during that short period of time. When the Sun comes up, the solar arrays start generating power again, and they become the primary means of power for the spacecraft. In addition, the arrays generate enough excess power to recharge the batteries back to full power, just like your car's alternator recharges your car's battery after it has been used to start the engine.

Nothing is perfect, of course, and solar arrays and batteries are no exception. The solar arrays, which are not energy efficient to begin with, become even less efficient over time, usually because they are degraded by radiation. At the end of their lifetime, solar cells may generate only half of the power they did at the beginning. The power requirements are probably the same, so the designer must overdesign the array to be twice as large as it needs to be at the start of a mission. This way, by the time the mission is over, the array can still produce enough power to get the job done. Solar power companies are developing new types of solar cells, such as those made with indium phosphide, which are more resistant to radiation than older-technology cells. Unfortunately, they are also more expensive to produce.

Solar arrays have other drawbacks, especially with larger spacecraft and when the spacecraft need to operate away from the Sun. In the latter case, solar intensity falls off very rapidly as one gets farther away from the Sun, necessitating either significantly larger solar arrays or significantly less power-hungry spacecraft. If we tried moving a spacecraft from Earth to Mars, we would have to at least double the size of the solar arrays just to be able to generate the same amount of power that we can get in Earth orbit. At the edge of the solar system, the Sun would appear like a bright star in the night sky and would not generate nearly enough energy to be of much use. For this reason, interplanetary spacecraft have to rely on some other means, as will interstellar spacecraft in the future. That other means is generally nuclear power.

As Einstein's famous formula $E = mc^2$ tells us, nuclear reactions, such as fission or fusion, could be used to convert mass into energy. The principle behind fission reactors, which are used extensively on Earth for power generation, can also be used to power spacecraft. All of the missions to the outer planets—Pioneer, Voyager, Galileo, and so on—have relied on nuclear power of a sort. Rather than carrying fission or fusion reactors, which would have been far too heavy to even get into space, these spacecraft generated power from the radioactive decay of plutonium in a device called a Radioisotope Thermoelectric Generator, or RTG. Regardless of the specific process, the great advantage of nuclear power is that you can get a lot of power out of a small amount of mass. The ideal would be a matter/antimatter engine that would convert mass completely into energy, and only a small amount of matter/antimatter (bundled in dilithium crystals, no doubt) could easily power a spaceship. So far, though, it is only a theory.

Regardless of how the power is produced, it ultimately must be distributed to the electrical equipment that needs it. The wiring in your house carries power from the junction box to the lights or the receptacles where you can use it. Similarly, the spacecraft must be "wired" to carry power from the power source to any electrical equipment, such as the spacecraft's computer, radios, and cameras. But the designer needs to decide *how* to distribute the power. Residential power in the United States is delivered at 120 volts. If you take your electric razor, designed to work at 120 volts, to Europe, you will need a special adapter to make it work since Europe uses a 240-volt system. There is no single standard for space power systems, either. Historically, many spacecraft have utilized 28-volt power supplies because their airplane predecessors did, but the International Space Station will utilize 160 volts. Higher voltages are nice because power that is distributed at higher voltages is more efficient; that is, less power is lost as heat in the transmission. The down side is that higher voltages are more dangerous: they are more likely to arc and damage people and equipment. Obviously, we feel happy in our ability to work safely around 120-volt systems on the ground. Keeping the voltage around that level in space as well gives us similar confidence.

In summer, it is not uncommon for heat-advisory warnings to be issued on sweltering days, suggesting we stay indoors. Likewise, in a winter snowstorm we tend to stay inside. Heating and ventilation engineers have helped society out by inventing air conditioning and central heating. We all enjoy climate control, and machines are no different. In summer, your car's radiator can overheat and explode; in winter, you add antifreeze so that ice won't form. Machines may be more tolerant of hot and cold extremes than humans, but they still need some degree of temperature control.

A good example of how critical temperature control can be is the unique Small Solar Probe I mentioned earlier. This spacecraft is designed to travel to within four solar radii of Sun center. At this point, the heat flux to the spacecraft is about 2900 times greater than on

Earth. A one-pound block of ice would vaporize into steam in about 0.01 second in this environment. To prevent itself from vaporizing at this distance, the spacecraft utilizes a unique multilayer heat shield, designed to prevent the heat load from reaching the spacecraft's instruments and electronics. If the heat shield has even so much as a pinhole-sized leak, due to a collision with a meteoroid for example, the spacecraft would melt within seconds. Fortunately, most spacecraft temperature requirements are far more forgiving.

There are three basic methods of controlling temperature. First, heat energy can be "radiated" from an object, just like the visible light that is being emitted from the Sun or the infrared heat energy that is being emitted by you and me; you wear a hat and gloves in the winter to stop heat from being radiated away from your hands and head. Second, heat energy can be "conducted" from one point to another directly through a material, as when you place a pot on a hot stove, when heat is conducted through the metal to heat the food. Finally, we can use "convection" to heat or cool things by using air to help bring heat in or take it away. This is the principle behind air conditioners or the heater in your automobile (unless you had one of the original Volkswagen Beetles). In space, since there is no air, convection won't work, so the standard Earth-based techniques of heating and cooling are useless except inside crew compartments. For this reason, designers have to take great care to ensure that the temperature of the interior compartments and the equipment areas remains adequate.

Spacecraft designers must balance the heat that comes into the spacecraft with the heat that escapes. If the heat input exceeds the heat output, the spacecraft will get warmer; if the heat output exceeds the heat input, the spacecraft will get colder. Light from the Sun will be absorbed and will be one source of heat, and even light reflected from the Earth will add to the heat load. More importantly, the spacecraft itself will generate heat because it is producing power. Your car's engine gets warm when you turn it on because some of the energy that is converted from the fuel into the mechanical energy required to run the engine is actually wasted in the form of heat. In other words, no power production technique is 100 percent efficient. Remember, solar arrays are considered efficient if they can convert only 20 percent of light from the Sun into energy. The other 80 percent is wasted, heating up the solar arrays so that they become quite warm, > 130° F, when illuminated. Thermal engineers must add up all of the heat energy that the spacecraft receives from outside, or generates itself, to find the total heat input.

The other side of the coin is calculating the heat output. Because people also have a temperature, about 98.6° F, they are radiating infrared light all the time. This warmth is what enables heat-sensitive night-vision devices to see us when it's dark. Because the spacecraft, like people, has a temperature, it too will radiate infrared light, and as light has energy, this radiation results in a heat loss. The spacecraft can also move heat around inside itself to help cool off warm areas and heat up cold areas. On crewed missions, this can be done with air; on uncrewed spacecraft, which probably would be unpressurized, the heat would have to be conducted—through heat pipes—from one point to another. Moving the heat around inside a spacecraft does not change the amount of heat that the spacecraft outputs, but it can help to regulate temperatures. This may seem like an odd concept at first, but your body does this kind of thermo-regulating all the time. When you're driving in the wintertime, your feet under the dashboard may get cold because the surrounding air is cold, yet on a clear day, your shoulders and head can get heated up by the Sun. So, while your toes are turning to ice, your head is sweating. All in all, you'd be a lot happier if you could move some heat from your head, to cool it down, to your feet, to warm them up. You could strap a heat pipe to your body to help make the transfer, but the easier solution is a bilevel control on your dashboard that can adjust the heating/cooling system to blow warm air on your feet and cool air on your head.

Often, the spacecraft designer will realize that it is not enough simply to move heat around inside a spacecraft. He or she must alter either the heat input or the heat output. Fortunately, in many cases this is simple to accomplish. Suppose the Sun shines on the street. A white sidewalk will feel much cooler than black asphalt because white reflects light from the Sun while black absorbs it. If the spacecraft wants to take in less heat, painting some outside surfaces white will decrease the amount of heat that is absorbed—much like putting white zinc oxide ointment on your nose in summer can prevent you from becoming sunburned. At the other extreme, spacecraft that need to take in more heat can be painted black to absorb as much light from the Sun as possible. Orienting the cool parts of a vehicle so that they point toward the Sun is a quick way to warm them up. In other cases, using an insulating blanket, just like you might use on your bed in winter, can help keep things warm. If these steps are not enough, more active measures, such as adding heaters, can be taken.

By now, it should be clear that the Sun is a critical element in the thermal design of a spacecraft. The Sun's heat load will change significantly as the spacecraft goes into eclipse, or leaves Earth orbit for an interplanetary cruise. The Sun heats the Earth to an average temperature of about 70° F, which is fortunate for us, because it is above the freezing point and below the boiling point of water. The range of temperatures on Earth provided the right conditions in which life could come into existence. The planets closer to our Sun are too hot to support life, while the planets farthest away are too

Plate 1. Buzz Aldrin, shown here on the Moon, was the Lunar Module pilot for *Apollo 11*. Neil Armstrong appears in pictures on the surface only as a reflection in Aldrin's visor. Armstrong carried the only camera. (Courtesy NASA)

Plate 2. The Space Transportation System (STS) is both a spacecraft and a launch vehicle. The spacecraft reaches orbit with the help of its rocket engines. (Courtesy NASA)

Plate 3. After its third servicing mission in December 1999, the Hubble Space Telescope snapped this picture of a cluster of galaxies called Abell 2218 in the constellation Draco, some 2 billion light-years from Earth. The cluster is so massive that its enormous gravitational field deflects light rays passing through it, providing a powerful "zoom lens" for viewing distant galaxies that could not normally be observed with the largest telescopes. (Courtesy AURA/STcI and NASA)

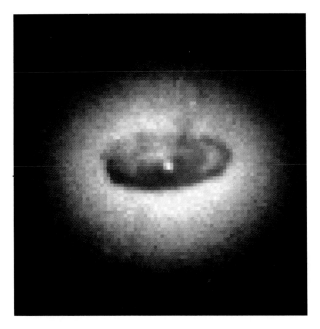

Plate 4. A picture of what is believed to be a black hole, being fed by a spiral-shaped disk of dust in galaxy NGC 4261, taken by the Hubble Space Telescope. (Courtesy AURA/STScI and NASA)

Plate 5. This image of Jupiter was taken from aboard the Hubble Space Telescope. The dark spot is the shadow of the inner moon, Io. (Courtesy STScI/JPL/NASA)

Plate 6. The glow seen near the engine pods of the Space Shuttle on STS-39 is due to an interaction with the neutral atmosphere. The aurora borealis, or northern lights, is seen in the background above the surface of the Earth. (Courtesy NASA)

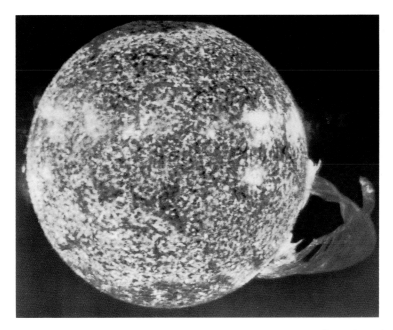

Plate 7. Solar flares, such as the one seen here, occur periodically. When they do, a large amount of mass is spewed forth into interplanetary space. If these particles strike a spacecraft, they become an additional form of radiation. (Courtesy NASA)

Plate 8. The Hubble Space Telescope's sharpest view of Mars. (Courtesy of AURA/STScI and NASA)

Plate 9. This *Landsat* 7 satellite image shows sections of Washington and Oregon. Mount St. Helens, in Washington State, is near the middle of the image. The city of Portland, Oregon, is near the confluence of the larger Columbia and smaller Willamette rivers in the middle third of the image. Snow-capped Mount Adams and Mount Hood can be seen to the right of Mount St. Helens and Portland, respectively. (Courtesy U.S. Geological Survey)

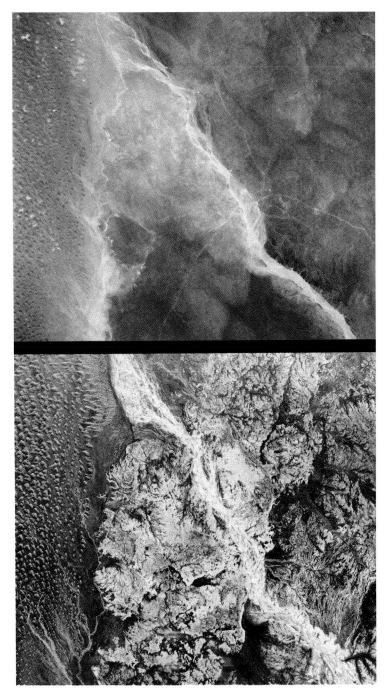

Plate 10. The Spaceborne imaging radar was used to study China's lost Silk Road. Seen in false-color radar (*left*) and visible light (*right*). The lost city of Ubar was rediscovered with the help of the radar images. (Courtesy NASA/JPL/California Institute of Technology)

Plate II. The planet Venus as seen by a radar image on the Magellan spacecraft. In visible light, the planet appears completely white because it is covered by thick clouds. (Courtesy NASA)

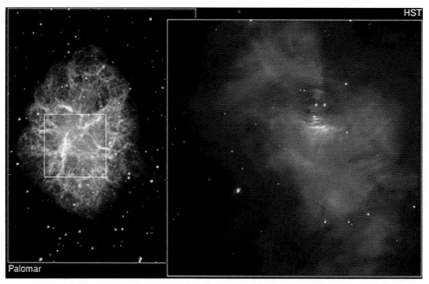

Plate 12. The Crab Nebula as seen by the Mount Palomar observatory (*left*) and the Hubble Space Telescope (*right*) in visible light. This is what remains of a star that exploded about 950 years ago. (Courtesy AURA/STScI and NASA)

cold. Of course, there may be exceptions. As Dr. McCoy of *Star Trek* was fond of saying, "It's life, Jim, but not as we know it."

In an aircraft, takeoff and landing are the two most important steps in the flight plan. With spacecraft, the launch and landing are equally critical. It takes the Shuttle's engines about eight and a half minutes to get it into orbit. During this critical time, if a problem develops with the engines the results could be catastrophic, as we saw all too clearly with the *Challenger* disaster. If an engine simply stops working, the spacecraft has no way to keep moving and may fall back into the ocean. Engines have stopped during a number of launches, but fortunately all of the missions were able to recover and continue. On the *Apollo 13* mission, during the firing of the second stage, one of the five engines mysteriously shut off. This dropped the thrust from the stage to 80 percent of its planned value, so the spacecraft was only being accelerated at 80 percent of the desired rate. Fortunately, the spacecraft was already high enough and moving fast enough that simply burning the remaining engines for a while longer solved the problem. On the Spacelab-2 mission, the Space Shuttle *Challenger* suffered a similar main engine cutoff when one of its three main engines quit during liftoff. After a few seconds of examining the situation, Mission Control came to the same conclusion, that burning the remaining two engines a little longer would get them to orbit, so they rapidly gave the command to continue with the mission. Unfortunately, a minor misunderstanding arose among the press corps attending the launch because they were not totally familiar with NASA's terminology. In every launch there is a very well defined orbit, characterized by an altitude and an inclination, to which the spacecraft is trying to get. There are an infinite variety of other possible orbits out there, but a particular one is chosen for each mission. With one main engine out, and operating on only two-thirds of the planned thrust, the Shuttle would not be able to reach the orbit of the original plan. Fortunately, it had enough thrust to get to "some" orbit, which is what NASA quickly ordered. The seldom-used command for this circumstance is "Abort to Orbit," which caused many reporters to panic and inform their respective networks with the breaking news that the Shuttle was about to "abort" and come crashing back to Earth. After several minutes of expecting the worst, the public was finally informed that the Shuttle was indeed in orbit and not in any immediate danger, so the news flashes quickly disappeared. Conversely, the mission-anomalies list that NASA keeps on every launch had only the briefest mention of the engine shutdown after the mission. It read simply: "Impact to Mission—None." Engine failures during landing are not so forgiving. The Mars Polar Lander was lost in December 1999 when its propulsion system switched off prematurely, causing the spacecraft to crash into the Martian surface.

Once a spacecraft is in orbit, it sometimes becomes desirable to change orbits. An excellent example is the Apollo Moon mission. The Apollo Command Module had to go into orbit around the Moon so that the Lunar Module could undock from it and land on the surface. However, after the Lunar Module returned its astronauts and cargo to the Command Module, the astronauts had to accelerate back to escape velocity for the trip back to Earth. To do this, the module needed a huge rocket motor instead of the small motors that might be part of the attitude determination and control subsystem.

Designing a propulsion subsystem for a spacecraft is similar to designing a mini launch vehicle into the craft itself. The spacecraft must have a rocket motor sized to give the right amount of thrust for the right amount of time, and enough fuel to go with it. Liquid-fueled propulsion systems are standard. Rocket fuel may be a single fluid (such as monomethylhydrazine), or it may be two separate fluids, both a fuel and an oxidizer. Two-propellant fuel systems are usually more powerful, as is the case when adding nitro to your gasoline tank to give it an extra push, but it requires additional hardware. The liquids must be stored in two separate tanks, which require two separate fuel lines to run the fuel through, and so on. Solid rocket motors, like the solid rocket boosters on the Shuttle, have the advantage of being even more powerful than their liquid counterparts. The down side is that solid motors can only be used once during a mission. Once they are turned on, they cannot be shut off. If you cut off the flow of fuel to a liquid motor, the result is the same as when your car runs out of gas: the motor stops immediately. The solid fuel is more like a charcoal briquette. You can douse the outside in water to cool it off, but since the briquette is on fire throughout its entire mass, in only a few seconds the internal heat will vaporize the water on the surface and the briquette is on fire again.

From ancient times, when sailors crossed the seven seas in wooden ships, the first piece of the vessel to be laid in place was the keel, the backbone around which the remaining structures are built. In like fashion, the spacecraft is built around its structural elements, just as the frame of a car provides a backbone for automobile designers. The structures themselves must provide enough strength to keep the spacecraft in one piece during launch or maneuvering, to contain an atmosphere for the crew, and for securing the other structures and subsystems. Furthermore, there has to be a way to lower the landing gear, extend solar arrays, and perform a variety of other tasks.

A minivan is bigger than a sports car. The size of an automobile is determined by the size of its engine, the number of passengers to be accommodated, the desired trunk space, and its intended use. The same is true for a spaceship, which has to contain all the subsystems, the crew, and the payload. Invariably, the subsystem designers and the crew want more space, just like you want more leg room and overhead cabin space on a plane trip, though the launch vehicle will be capable of carrying only a finite amount of mass crammed into a finite amount of space. The spacecraft must be light enough for the launch vehicle to be able to propel it into orbit, and it must be small enough to fit on top of the launcher.

One way of limiting mass is through choice of materials. Steel is a good choice for constructing large office buildings because of its great strength, but you would not want to drop a steel girder on your toes. Spacecraft designers like to use lighter-weight materials that provide higher strength. Aluminum is the preferred choice if solid metals are used, but even solid aluminum is often too heavy. Significant weight savings can be found by building panels out of aluminum honeycomb material. The honeycomb is far lighter than solid material, yet it is incredibly strong, just like corrugated cardboard is stronger than ordinary cardboard of the same thickness. Using composite materials such as graphite epoxy, the stuff of which many tennis rackets and golf clubs are made, can reduce the weight even further. As with all things, there are advantages and disadvantages to each option. The graphite epoxy materials are lighter, but they are more costly to purchase and assemble. In some cases, they may even absorb water from the Earth's atmosphere, which may cause them to expand and thus cause misalignments in places where tight tolerances (such as lining up two screw holes on different panels) are required. The art of engineering is to consider the pros and cons of cost, mass, ease of assembly, and strength, then arrive at a solution that can please everyone.

In addition to providing strength, the structures must also provide rigidity, so that the spacecraft does not vibrate excessively. The launch vehicle itself will be loud and shaky, and the spacecraft will be subjected to a great deal of stress even during a short launch.

Once in orbit, the minor forces that pose a problem for the attitude determination and control subsystem will also have an effect on the structures. A spacecraft that rotates rapidly must be designed with the proper amount of strength and stiffness so that it doesn't shake itself apart.

Mechanisms must be designed so that they can operate in the vacuum of space. This requirement limits the choice of materials and makes the use of most lubricants impossible. Typical greases that you might use on your automobile would literally boil away if placed in a vacuum, becoming a significant source of self-contamination. High reliability is a necessity, as the failure of even the smallest mechanisms can be disastrous for a spacecraft.

A radio is the spacecraft's lifeline to the outside world. Even if it does not have a crew, a spacecraft must be able to receive signals from Earth so that it knows what to do. In return, the spacecraft must also send signals back to Earth so that ground controllers know where it is, where it's pointing, and whether its most recent task has been accomplished. Every spacecraft routinely sends back a small stream of data that describes its health: how much power is available, at what temperature are the critical components, how fast are the astronauts' hearts beating, and so forth. This gives ground controllers an idea of how well the various subsystems are functioning and also helps them solve problems if something goes wrong.

A critical yet often overlooked element of the communications subsystem is the antenna. If the local car wash snaps the antenna off your car during the rinse cycle, your radio reception stops. Similarly, if the spacecraft's antenna doesn't work as planned, big problems can result. A good example is the Galileo probe to Jupiter. The mission was severely limited when its high-gain antenna refused to deploy (fig. 24), keeping the spacecraft from transmitting data back to Earth at the desired rate. As a result, a great deal of the information

Figure 24. The Galileo spacecraft before launch. The high gain antenna, which looks like a large umbrella near the top, failed to deploy after launch. Note the size of the spacecraft relative to the technician near the bottom. (Courtesy NASA)

that could have been gathered by the spacecraft was lost. Fortunately, its smaller antenna is working fine, and Galileo was able to send back pictures of the fragments of comet Shoemaker-Levy that impacted Jupiter's night side in 1994 (fig. 25). A Russian Mars probe suffered a completely different, and totally fatal, antenna failure when a faulty command told the spacecraft to turn its antenna away from the Earth. Years later, the spacecraft may still be patiently waiting for further instructions from Earth that it will never be able to hear.

If we master the nontrivial tasks of deploying the antenna and pointing it toward Earth, we must also ensure that the communica-

Figure 25. Comet Shoemaker-Levy impacted Jupiter with the force of a detonated nuclear weapon. The impact occurred on the far side of Jupiter, as seen from the Earth, but was witnessed by the Galileo spacecraft. From left to right: before impact, start, maximum, and end. (Courtesy NASA)

tions subsystem can manage all of the radio traffic expected. The radio traffic, or more appropriately the amount of data to be transmitted and received, is determined almost entirely by the payload. Remote-sensing spacecraft must transmit to Earth thousands of electronic images captured by their cameras. Communications spacecraft may have to transmit and receive thousands of separate phone calls simultaneously. Other instruments may have to deal with completely different kinds of data, but all of the data have to be accommodated.

The amount of data to be transmitted is certainly the primary factor in determining how to build a functional communications system, but it is not the only one. Being farther from the Earth requires the spacecraft to use more power so that the signal will be strong enough to be received on Earth. At the same time, spacecraft that travel far from Earth must carry along larger antennas so they can receive the signals being broadcast to them. Just like tuning your car's radio to AM or FM, if the spacecraft isn't listening at the right frequency it won't hear the signals being sent to it. Selection of the frequency, in order to be compatible with ground stations like the Space Ground Link System or space relay stations like the Tracking and Data Relay Satellite System, determines the hardware that must be used to generate, and decipher, the radio signals in question.

Designing and manufacturing communications for a starship would not be a trivial task, but it is probably possible even with today's technology. Personal communicators are already part of today's world. Today's cellular phones are only slightly larger than the chest lapel communicators depicted in science fiction film. The flip top on *Star Trek*'s original communicators is strikingly similar to the flip top on today's cell phones. The only new challenge the Enterprise would face would be the vastly different communications standards and protocols utilized by the different civilizations they encounter. On Earth, the use of certain frequencies for certain functions is enforced by treaty. Some frequency bands are used for AM radio, others for FM radio, others for TV, others for satellite communications, and so on. A lot of radio frequency noise would be generated by an advanced civilization, and a starship's communications system would have to be able to lock onto signals in a wide variety of frequency ranges. Undoubtedly, some frequencies would be reserved for future spacecraft communications. Why else would Captain Kirk make repeated use of the phrase, "Open a hailing frequency, Lt. Uhura."

A hard lesson in choosing correct communications frequencies was learned after the launch of a Thor-Agena rocket in the early days of the U.S. space program. After the rocket completed half an orbit, ground controllers sent the command to activate a satellite that had just been jettisoned from the rocket. A few minutes later, radar on the ground detected not only a rocket and satellite, but over fifty smaller pieces of debris. On further investigation, engineers realized that the frequency that turned on the satellite also activated the rocket's self-destruct mechanism, causing it to explode. Fortunately, the satellite survived undamaged, but there were a lot of shocked faces around the room at the command center. Earlier there had been a debate about turning the satellite on before launch, which would have caused the rocket to explode on the pad.

To avoid problems like this, many of the radio signals we use today are encoded with the specialized styles of their originators. Whether we are transmitting our credit card number via the Internet or sending commands to our satellite, we do not want anyone intercepting this information and being able to misuse it later. Most countries would be greatly distressed if other countries learned how to turn off spy satellites as they were passing overhead. Deciphering encrypted transmissions, even after they have been located, is usually not a trivial task. Many codes can be broken once we know the "key." For example, you and I could use this book as a key for our own special code. I could send you a message that, if intercepted, would look merely like a random string of numbers. However, since you possess the key, you know that each number refers to a specific page, paragraph, word, and letter in this book. Knowing the key, you can then decipher the code and translate the seemingly random numbers into letters and words. Modern codes often use completely random keys, such as the radio static hiss of the Earth's atmosphere, as their basis for encryption, making it virtually impossible—or so spacecraft operators hope—to break the code. Strictly speaking, however, the task of deciphering the encoded messages sent to it is the job of the spacecraft's computer.

Like a communications system, a computer is an integral part of every spacecraft. The computer deciphers commands sent to it by the ground and distributes and receives data to and from the other spacecraft subsystems. As such, it is often called the "command

and data handling subsystem." Although these functions are fairly independent, it is convenient to utilize a single computer for both functions.

Spacecraft computers were lampooned from the start. In the movie *2001*, a malfunctioning HAL computer (which got its name by choosing the letter in the alphabet that preceded each of the letters IBM) tried to override its human controllers, and even send them to their deaths, in an attempt to ensure the mission continued as it saw fit. Knowing what a computer can be trusted with, and what a human crew should handle, is often the hardest problem in computer design. The simple word processor I am using to type this book does a fine job of capturing the text and editing it, but the task of composing the words is left entirely up to me.

The human brain is a far more complex computer than we have ever been able to build. We could, and did, build spacecraft capable of traveling to the Moon, gathering up some soil samples, and returning them to Earth. However, a critical element in assigning a computer to execute this task is knowing what rock it should pick up. This was the reason for selecting astronauts with a strong science background for some of the Apollo missions. People with experience in field geology, like Harrison Schmitt on *Apollo 17*, can do a much better job of deciding what rocks or soil samples would be most useful to scientists back on Earth because the human brain is capable of analyzing complex data in very short periods of time.

For example, suppose you are driving down the road and hear the words "pull over" blaring through the open window of your car. You look in the rear-view window and see a police car, driven by a uniformed officer, so you pull over. Within a few seconds you have decoded the data, the verbal command and the image of the police car, validated that it comes from a police officer and not a prankster, and executed the instructions. A spacecraft must be programmed to do the same thing—decode, validate, and execute—every command it receives from the ground.

The design of the spacecraft's command and data handling subsystem is not too different from designing a computer and data transmission network on the ground. For example, my PC must be capable of deciphering commands from my keyboard or my mouse, and it must also be able to read in data from my scanner, my disk drives, my CD-ROM, and even the Internet. A spacecraft's computer will need to decipher commands received through the spacecraft antenna, and it must be able to read in data from the attitude determination and control sensors, temperature sensors, and so forth. The key differences are that the spacecraft computer must be designed to work in space, which is a very different environment from Earth, and it must also be capable of handling a wide variety of problems on its own without a helping hand to restart it if something goes wrong.

Lightning on Earth can cause power surges that can damage our ground-based PCs. For this reason, we often use surge protectors to prevent these power spikes from reaching our computer. In space, even if it is located deep inside the spacecraft and apparently protected from the elements, the spacecraft's computer will still be pelted with radiation from space, which can have a big effect on the performance of even the simplest electronic hardware. Unfortunately, the radiation is so energetic that there is no way to prevent it from reaching the computer, so the computer has to be able to deal with it.

Information in computers is stored as binary information, with all data appearing as some combination of ones and zeros. Just as there is no such thing as being a little bit pregnant, a data bit is either "on" (one), or "off" (zero), with nothing in between. The problem is that radiation is capable of switching bits from one to zero or vice versa.

Because electronic devices are so small, with dimensions measured in micro-inches, even a single energetic particle can create a large, localized electric field in the device that can cause it to upset momentarily, or fail permanently. If you've stored information about your reentry trajectory in your computer as a series of ones and zeros, you want to be certain that the information stayed correct until you need to use it. One wrong digit, and a safe landing at Edwards Air Force Base can turn into an unexpected visit to Disneyland. Unfortunately, the radiation environment can cause the digits to change. Asking a computer to determine when it has made an error is like asking schizophrenics to determine when they need psychiatric help. It is not easy, but it is essential that the computer be able to know when it has made a mistake *and* be able to fix it. This is called *error detection and correction*.

An error detection and correction capability comes at a price: your computer systems are much more complex. If you have a piece of information stored in memory as a series of ones and zeros, and suddenly a zero flips to a one, how will you even know if something changed? One way to do this is to add another digit, called a *check bit*, to the end of each string of numbers. This is an ingenious approach: if the number of ones in the original word is odd, the check bit is one; otherwise it is zero (fig. 26). As the spacecraft computer pulls each word out of memory, it recomputes what the check bit ought to be and compares this value to what it really is. If these values are different, we can be certain that an error has occurred. Unfortunately, knowing that an error has occurred is only the first part of the problem—error detection. We don't yet know exactly what the error is. The problem may have occurred in the check bit itself, so that the word is really correct. The only way to determine where the error did occur is to store each word in two separate locations, call them columns A and B, then compare the answers later.

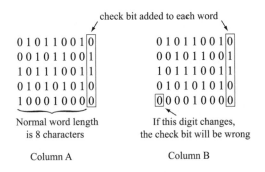

check bit added to each word

0 1 0 1 1 0 0 1 0	0 1 0 1 1 0 0 1 0
0 0 1 0 1 1 0 0 1	0 0 1 0 1 1 0 0 1
1 0 1 1 1 0 0 1 1	1 0 1 1 1 0 0 1 1
0 1 0 1 0 1 0 1 0	0 1 0 1 0 1 0 1 0
1 0 0 0 1 0 0 0 0	0 0 0 0 1 0 0 0 0

Normal word length   If this digit changes,
is 8 characters   the check bit will be wrong

Column A   Column B

Figure 26. Error detection and correction can be done by adding a check bit to each word. The check bit is one if the number of ones in the word is odd, and it is zero if the number of ones is even. If a bit flips at a later time, the check bit will be wrong.

If the word from column A is right, and the word from column B is wrong, then we know which word is incorrect. By comparing the wrong answer in column B to the right answer in column A, we know exactly where the error is. Once we know which number got flipped, we can reset it to the correct value and continue. The problem is that all this takes extra storage space and extra time, which slows down the computational process. As a result, many spacecraft processors are less robust than your standard desktop PC. There is far more computing power available in a new PC than was available on the Apollo spacecraft.

Even if we come to grips with error detection and correction, we still have to deal with the problems associated with the great distances involved in interplanetary, let alone interstellar, travel. Remember that it takes a radio signal from the Earth about one second to reach the Moon. When the Apollo landers touched down on the Moon, we knew about it here on Earth one second later when the radio signal reached us. One of the trickier things attempted with the last Apollo landing was a live broadcast of the Lunar Module's blastoff from the Moon. On earlier missions a camera had been fixed on the lander, but the camera couldn't move. After the Lunar Module had reached an altitude of only 20 or 30 feet, the camera couldn't see it any more. On *Apollo 17*, the camera had the capability to pan up and follow the liftoff. The trick was that the camera had to be controlled manually from the ground. It took one second for the command sent to the camera to be received on the Moon, then another second for the response to be received on the ground. That meant there was a two-second delay between the time the ground controller did something and the camera showed if it worked. The operator had to listen to the liftoff countdown, bear in mind that what he was hearing was one second behind the actual time on the Moon, and then start panning the camera up two seconds before liftoff. Since he couldn't see the result until after the fact, he had to hope that he panned up at the right time and the right speed. Fortunately, everything worked fine, and we have a nice video of the last blast-off from the surface of the Moon (fig. 27).

If a spacecraft has a problem and help is only one second away, as in the case of the *Apollo 13* emergency, our chances of finding a solution are pretty good. However, recall that when the Mars Pathfinder touched down on the surface of Mars in 1997, it was about twenty minutes before that fact was confirmed on Earth. This means that it took at least forty minutes to learn of a problem and send up a possible solution. That makes the possibility of helping out from Earth very remote. Spacecraft can't always wait for help to come from the Earth, so they have to be programmed to deal with all kinds of emergencies on their own. This ability to operate independently of ground controllers is called *autonomy* and must be built into the spacecraft's computer systems. This is not easily done, and it explains why Gene Kranz, Apollo flight director, said "unmanned missions are the toughest ones."

Figure 27. The only complete video of a takeoff from the lunar surface was captured remotely during the *Apollo 17* blastoff. (Courtesy NASA)

A significant part of autonomy is the extensive programming that the spacecraft must carry along to tell it what to do under various circumstances. Intricate "if this happens, then do that" routines are needed to deal with all possible failures. For example, if a problem develops with the attitude determination and control system and the spacecraft loses the ability to keep its antenna pointing at the Earth, it will not be able to communicate with the ground. The spacecraft will have to sense that it has lost contact with the ground and then search for the Earth and relocate it on its own.

The problem with relying on extensive if-then statements (which computer scientists will tell you is sloppy programming, anyway) is that we can't always predict every possible failure. A few years back, the control systems at a dam not too far from my house sent the command to open the floodgates and leave them open. A few hours later, a lot of unhappy boat owners found their prized possessions sitting in mud because all of the water had run downstream. The problem, which was later easily corrected, was that the control system simply did not know what to do when it encountered the particular combination of inputs it received, so it decided to open the doors and wait for further instructions. It might seem like an odd course of action for a computer to take, but I know I do similar things all the time. While at the grocery store just the other day, my wife gave me explicit instructions to "go get a 16-ounce bag of frozen, crinkle cut, name-brand French fries." As luck would have it, I could find 16-ounce bags of fries, I could find crinkle cut fries, and I could find name brand fries, but I could not find a single bag that fit all of the criteria. So I did what any self-respecting husband would do: I sent my son back to my wife to ask for further instructions.

Since problems like this can, and do, occur, it would be nice if computers, and husbands, could simply "think" for themselves. That's the principle behind the field of study known as *artificial intelligence*, which seeks to determine how we can get computers to think. The first challenge is to define what thinking really is. Often it is very difficult for us to articulate exactly how we came to a conclusion, let alone write down specific instructions that would enable a computer to get the same answer. That is why we have to examine the reasoning processes that we, or computers, use to determine information. Sometimes we can use logic, often we have to resort to probability or statistics. In still other cases, we may be tempted to use what is known as *fuzzy logic*, a method of reasoning that allows two different answers to be partially correct at the same time.

Two more centuries of computer development at today's pace could result in quantum computers that are billions and billions of times more capable than today's PCs. When coupled with more advanced software, the results would be a computer system that is almost infinitely capable, the fulfillment of the *Encyclopedia Galactica* from Isaac Asimov's *Foundation* series. Tomorrow's *Enterprise* by itself would surely contain more computing power than is available throughout all of Earth today.

Once you are in space, the problem of figuring out where you are, and how to get from there to where you want to be, is very complicated. On the surface of the Earth, land vehicles can follow roads or landmarks during their travels. Surface ships or aircraft can navigate effectively by compass or by using the Global Positioning System (GPS) navigational constellation of satellites. In space, especially interplanetary or interstellar space, there are no landmarks to use other than the stars and planets themselves. Fortunately, we have been navigating by the stars for thousands of years. Over two thousand years ago, the three Wise Men were guided to a manger in the town of Bethlehem by simply following a star. Every Boy Scout or Girl Scout can easily locate the North Star and find the four points on the compass from it. In fact, a quick measurement of the angle between the North Star and the horizon gives you a good estimate of your latitude—the distance north of the equator. In the Southern Hemisphere, we would have to use different stars, but the principle remains the same. Because the stars, as opposed to the planets that "wander," are fixed in the heavens, we can measure changes in our location on the Earth by observing changes in the appearance of the stars. Latitude is the easy part. Astrolabes have been used for centuries to help determine latitude. The trickier part is determining longitude, your east-west position on the Earth. Again, the stars can help here. If we measure exactly when certain stars pass directly overhead and we know the time of year, we can deduce our longitude as well. Using nothing but the stars and some fairly primitive measuring devices, along with some prior knowledge of what the stars are supposed to look like, we can easily determine our position on the surface of the Earth. Determining where you are is the first step in navigating successfully to your destination. The trick is that navigating in space means being able to determine not only your latitude and longitude, but also your altitude.

A spacecraft's motion is often directly measured by observers on the ground. By tracking radio signals broadcast from the spacecraft itself or by using radar returns, the ground station can determine the spacecraft's position and velocity. Lower-orbiting spacecraft can use the GPS constellation for navigational information. Ultimately, though, we'd like the spacecraft to be able to figure out its position on its own. Information gathered by the Earth, Sun, or star sensors from the attitude determination and control system can be used to infer not only orientation, but also position.

The need for an accurate estimate of your position actually gets more critical as we move away from Earth and into interplanetary space. Near the Earth, we usually worry that we might drop too low in our orbit and reenter the atmosphere too quickly, causing us to land in the wrong place. In interplanetary space, the chances of hitting something are so remote that the reverse situation becomes the problem. Instead of trying to figure out how not to run into planet Earth, we have the difficult problem of trying to run into planet Mars. The red planet is about 48 million miles away from Earth. If we leave Earth orbit to head for Mars and our route is off by as much as one inch over a distance of 100 yards, we'll miss the planet Mars entirely.

Fortunately, nature provides some of its own interstellar beacons as we move farther from our own solar system. Using a radio telescope while a graduate student at Cambridge University in 1967, Jocelyn Bell detected a radio signal coming from the constellation of Vuplecula with regularly timed pulses, repeating every 1.33728 seconds. The strong, regular signals made astronomers think, at first, that they had found extraterrestrials trying to communicate with us; therefore, they denoted the signal with the acronym LGM—little green men. In time, hundreds of similar signals were also discovered, and we realized that these signals emanate from a special kind of star. These special stars, which we now call pulsars, are superdense stars that sometimes form when a large star implodes, crushing the central region and then blasting the outer regions into interstellar space in an event called a *supernova*. Pulsars have strong magnetic fields and send out strong radiowave blips from their north and south poles. The stars spin rapidly, so they resemble radiowave lighthouses. Because each pulsar emits a unique radio signal, these signals can be used as navigational beacons on interstellar trips.

Spacecraft carrying a living crew have to have a hospitable environment for them to live and work in. The temperature must be maintained in the comfortable range. In addition, the atmosphere must remain breathable, so there must be some means of introducing fresh oxygen into the air and removing stale air, poisoned by exhaled carbon dioxide or other impurities. As simple as this is on the surface, it's a little trickier in zero-g.

In the absence of gravity, dust and debris does not drop to the floor; it floats right there in front of you. A piece of cereal or a spec of paper could easily be inhaled and cause serious problems to a crew member. To prevent these problems, spacecraft must be kept incredibly clean. On *Gemini 3*, astronaut John Young pulled out a sandwich instead of the incredibly edible space rations specially prepared for zero-g, and handed it to his commander, Gus Grissom. Grissom took one bite so the astronauts could make their point about missing Mom's home cooking, but he quickly put it away. The bread would have rapidly crumbled and sent tiny specs of food floating into the air for the astronauts to breathe in later. Without gravity, not even the air itself wants to move. Fans must be used to circulate the air inside the spacecraft. If not forcefully circulated, the carbon dioxide that is exhaled by a crew member won't go any farther than out the crew member's nose. It will sit there and be inhaled again. Astronauts have been known to get headaches from inhaling too much carbon dioxide if they happen to fall asleep in a spot of dead air. All things considered, life in space is quite different because of the absence of gravity.

Jumping off your chair gets you a fraction of a second of zero-g. Going into space gets you continual zero-g. When the body is exposed to zero-g, funny things begin to happen. First of all, your stomach—which is accustomed to having gravity pull it down toward your feet—rises up. That is the queasy feeling you get when you go on a roller-coaster, and it is also the source of "space sickness" for a high percentage of astronauts. Second, your heart is accustomed to having to pump blood up against gravity to get it to your head. When gravity is taken away, the heart pumps too much blood to the head, so your face looks a little swollen and puffy. The brain registers this abundance of fluid and sends the message out to dump it. So you head for the bathroom, even if you don't need to go. Astronauts can easily get dehydrated, so they have to force themselves to keep drinking water all the time.

Another long-term consequence is muscle and bone loss. If you want to go to the refrigerator in your house and get something to drink, you have to expend energy to stand up and walk across the room. In zero-g, you don't have to expend any energy to stand up, since there is no gravity to fight. Pushing yourself off the wall with your little finger will provide you with enough velocity to coast across the room. Because your body doesn't have to work hard, it doesn't. It gets accustomed to not having to fight against gravity all the time, and you become weaker. Astronauts and cosmonauts returning from many months in space are often too physically weak to even walk once they're back in the Earth's gravity. They must be carried to the ambulance for the drive to the hospital, where they start their recuperation. Along with the muscle loss, your bones lose minerals and become more brittle. Dr. Jerry Lininger—the U.S. astronaut who spent five months aboard the Mir Space Station—reported a bone loss of about 20 percent from his hips. After a year back on the Earth, he had only regained half of what was lost. Although he managed to walk off the Space Shuttle, Dr. Lininger described stepping over his chair as being "like climbing Mount Everest." This physical degradation would be a definite detriment to a planetary mission using conventional technology. It would be very difficult to justify sending a crew on a nine-month mission to Mars knowing that the crew would be too weak to do anything but rest for several days after they got there.

Unfortunately, there is no simple way for a spacecraft to produce artificial gravity. Gravity is a fundamental force between two objects having mass. The only way to simulate it would be through other forces. The spacecraft could fire its propulsion system, which would accelerate the spacecraft and have the effect of providing a 1-g environment for the crew. However, this results in a rapid use of fuel that is just not practical for today's spacecraft. One method that is possible is to simply rotate the vehicle, just like in Arthur C. Clarke's *2001: A Space Odyssey.*

As we already saw, to rotate the Starship *Enterprise* just once per minute, the outer hull would have to move at a speed of about 40 mph. This circular rotation would exert a centripetal acceleration on the hull and the crew members inside it that would cause the crew to feel an outward acceleration. In this case, the acceleration would be only about one third of a g, but the principle behind this rotational motion could be used to simulate gravity. One concept being examined today is to utilize rotating beds, so that the occupant would feel a force pulling him or her downward toward their feet. Although it is not induced by gravity, this force can imitate its effects and—we hope—help to ward off the mineral and muscle loss that extended periods of weightlessness can produce.

If we manage to provide our crew with the means to overcome the effects of weightlessness, and if we can provide them with a self-sufficient way to produce food to eat and air to breathe, we're almost ready to venture into deep space. One last concern we have to address is the possibility of space travelers encountering a unique strain of space-borne germs. As shown in the movie *The Andromeda Strain*, our human bodies would be defenseless against new bacteria. Even on Earth we see this quite clearly. Individuals who volunteer to spend a year at the Antarctic research station on the South Pole enjoy several months of virtual disease-free existence because the frigid temperatures prohibit the growth of any new infections. However, when the researchers return to warmer temperatures, they are virtually guaranteed to get sick. It is not that their bodies have become weakened by their stay, but that they have not been exposed to any of the newer strains of bacteria that are constantly being formed. Our limited exposure to these new strains on a day-to-day basis allows us to build up immunity. However, when we have no exposure for several months, our bodies have nothing to build up immunity to. Just like the Martians in H. G. Wells's *The War of the Worlds*, who perished a few days after invading the Earth because they had no resistance to our diseases, space travelers must take great care not to expose themselves to alien life forms. Undertaking a mission "to seek out new life and new civilizations" could ultimately prove quite fatal—both for ourselves and our newfound friends.

If we protected our astronauts by keeping them inside a spacecraft, they would notice little difference as they traveled to a space station, the Moon, the other planets, and beyond. However, the spacecraft and its equipment—which are exposed to "space"—would notice some fairly significant differences between the kind of environment we live in on the Earth's surface and the kind of environment found out there. To understand what space is really made of, and to appreciate the challenges that working in space pose, we first need to understand a little about the Earth's atmosphere.

Out of This World:
The Environment of Space

*Toto, I have a funny feeling we're
not in Kansas anymore.*

Every week our television reminds us that space is the final frontier. This philosophical definition is undoubtedly correct, but it does not paint much of a picture about what's out there. Many of us picture space as an astronomer's playground. A picturesque conglomeration of black holes, white dwarfs, nebulas, supernovas, and star clusters punctuated with random bits of flotsam and jetsam found in asteroid fields. Countless centuries of wondering, and over forty years of observing actual space travel, have acquainted most of us with some concept of "space," but what do we really find when we get there?

The first thing we notice is that the Earth's atmosphere isn't there. This may seem obvious, but it is also important because the atmosphere keeps us alive. It contains the air we breathe, and it helps protect us from high-energy particles from space such as solar protons and galactic cosmic rays that could harm us. Our atmosphere is mainly made up of nitrogen (78 percent) and oxygen (21 percent), with a few other assorted elements like argon, neon, and helium thrown in for good measure. The atmosphere also contains varying amounts of water vapor and pollutants that are emitted from automobiles, industrial plants, and so on. In addition, the Earth's atmosphere is home to various weather phenomena, from tornadoes and hurricanes to sandstorms, which can bring with them widespread devastation of crops and property damage. Weather is not unique to Earth. Jupiter's great red spot is an example of a great hurricane-like storm on that planet (plate 5).

On the surface of the Earth, the worldwide average atmospheric temperature is about 70° F, which feels quite comfortable to us. If the temperature were much hotter, or much colder, on average, life as we know it might never have come into existence.

On Earth's surface, a 2-liter sample of Earth's atmosphere (the size of a large soda bottle) has a weight of about 0.0054 pounds. As we move up in altitude toward space, we find that the air gets less dense, so a 2-liter sample would weigh less. A 2-liter sample of air in Denver at the Mile High Stadium would weigh a fraction less than it would in Miami. By the time we got to the International Space Station, a 2-liter bottle of air would weigh less than 0.000,000,000,004 pounds—over one billion times less than it does on Earth's surface.

Most adults have a head that's about 6 inches in diameter. The column of air that sits on the top of your head and reaches all the way to space would have a weight of about 400 pounds. We certainly don't feel like we're carrying around an extra 400 pounds when we try to stand up because the pressure from the air is with us all the time and pushes us equally from all sides, up and down, so it just feels normal to us. But when we try to build machines to travel from down here to up there, this pressure becomes important. To verify that the atmosphere really does have a weight we can use a simple device called a *barometer* (fig. 28). We simply fill a glass tube with a liquid, usually mercury, and bend it as shown. The mercury in the

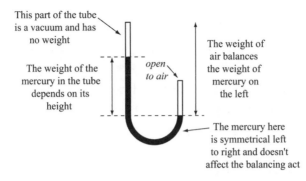

This part of the tube is a vacuum and has no weight

The weight of the mercury in the tube depends on its height

open to air

The weight of air balances the weight of mercury on the left

The mercury here is symmetrical left to right and doesn't affect the balancing act

Figure 28. A simple device called a barometer can measure pressure that is due to the atmosphere. Near sea level, the atmosphere exerts a weight of 14.7 pounds over every square inch of area.

U-shaped part of the tube exerts just as much force on the left as it does on the right, so it doesn't really contribute to the final answer. Just like a set of scales, the mercury in the left part of the tube will automatically adjust itself so that its weight is equal to the weight of the column of air on the right. If the weight of the air is greater, the air will push more mercury into the left part of the tube until the weights balance, and vice versa. By measuring the height of the column of mercury with a ruler, we can indirectly measure the pressure due to the weight of the atmosphere. Average atmospheric pressure would support a column of mercury about 30 inches high. Barometers are routinely used in weather forecasting. Lower pressures typically mean bad weather is approaching, while higher pressures forecast clear skies.

Scientists, and others, have developed the unfortunate habit of mixing terms in their conversation. I just described pressure in terms of "inches of mercury," when pressure is really a measure of weight per unit area, not distance; I have also quoted some values in English units, like miles, and given some values in metric units, like liters. In 1999, the $125 million Mars Climate Orbiter was lost as the spacecraft approached Mars due to a mixup about metric and English units. Two different teams working on the probe used different measurement systems. The spacecraft was given the wrong numerical command, a number that was correct in the "right" system of units, but which caused the spacecraft to fly too close to the planet and burn up in the Martian atmosphere in the "wrong" system of units. Obviously, choosing a single system of units and using those units consistently is important, so it makes sense to get them right before we go any further.

Americans are accustomed to using English units such as inches, feet, ounces, and pounds. Most English units are familiar to us, but some are not. For example, the English unit of measure for mass is the "slug." Since the slug is not a common term, engineers sometimes take a shortcut and invent units of convenience, like "pounds of mass," to make the math simpler. If we try to use the metric system of units, we can still have problems. We often speak of weight in terms of kilograms, but the kilogram is really a measure of mass. If something weighs 10 kilograms, we should really say that it has the weight that an object with a mass of 10 kilograms would have. The weight of an object would be its mass, in kilograms, multiplied by the acceleration of the Earth's gravitational field, in meters per second squared. The metric unit for force is the "Newton," abbreviated simply N and named after the gentleman we mentioned in an earlier chapter.

This may seem like we're splitting hairs, but as we saw in the case of the Mars Climate Orbiter it is a very important issue. Most scientists use the metric system, and the *Star Trek* of the twenty-third century uses it as well. As luck would have it, Americans and their space program still use mostly English units. When I was in the eighth grade, I was first introduced to the metric system and fought it with a passion. Why bother learning metric when we've already got a perfectly good system of units in place? By now, I've swung to the other side of the pendulum. As a scientist, I use only the metric system of units in my calculations for one selfish reason—this is an easier system. It's far easier for me to remember that it takes 1 calorie of heat to raise the temperature of 1 gram of water by 1° C, than it is to remember that it takes 0.000,95 BTU of heat to raise the temperature of 0.035 ounces of water by 1.8° F. Yet, unless we are doing scientific calculations, we may not have much motivation to change. We should bear in mind, however, that virtually every country outside of the United States uses metric units. Even England doesn't use English units any more. As the global economy matures, the United States will have to produce products in metric units to satisfy customers in other lands. In the meantime, though, we still manufacture most of our goods in English units. It's far easier to go down to the local hardware store and buy a 2 by 4 inch piece of lumber, than to try to get a piece that measures 5 by 10 cm.

As the Mars Climate Orbiter example shows, the United States is slowly converting to metric. On the surface one might think that we could make the conversion to metric within a generation or so if we made a conscious effort. However, history shows that it is far harder to change a tradition than you might think. For example, the solid rocket boosters on the Space Shuttle are manufactured in Utah and have to be shipped by train to the launch site in Florida. In order to fit through a tunnel in the mountains, the boosters can't be much wider than a set of train tracks. The U.S. standard railroad gauge, the distance between the rails, is an odd 4 feet 8.5 inches. Why was that gauge used? Because that's the gauge they used in England about two hundred years ago, and English expatriates, who were familiar with the English standards, built the U.S. railroads. Why would the English use this gauge? Because it is the same size as the pre-railroad tramways, and the people who used to build the tramways also went on to build the railroads. Why did the tramways use this gauge?

Because the people who used to make wagons also made the tramways, and they found it convenient to make them the same size. So why were the wagon wheels spaced an odd 4 feet 8.5 inches apart? Well, the new wagons had to fit in the ruts made by the old wagons, or they would give their passengers a very uncomfortable ride. And who made the first wagon wheel ruts in England? The Roman legions, two thousand years ago. So there you have it: the size of the Shuttle rocket boosters may ultimately be limited by the size of wagons built in Imperial Rome. Once you get used to building something of a fixed size, or on a fixed system of units, the repercussions can last for centuries, and even millennia. So rather than allow future generations to become confused by our lack of precision, let us make sure we know exactly what we mean by certain terms, and let's begin by returning to our discussion of pressure.

Since pressure is a measure of weight per unit area, we need to make sure we understand what weight means before we can understand pressure. To understand weight, we first have to understand mass. Mass is a fundamental property of matter. It is something so basic that we really can't describe it in any simpler terms. In some sense, it measures the amount of "stuff" of which something is made. You and I have mass, as does the Earth and almost everything else around us. Light or electric and magnetic fields are about the only things we find in nature that have no mass. As we saw earlier, when any mass is left at the mercy of a force it will be accelerated. By definition, the force that our mass feels due to gravity is our weight—on the Earth. If we travel to another planet, where the gravitational force is different, our mass will be the same but our weight will be different because the acceleration due to gravity is different. If we could go someplace where there is no gravity at all, then we would not have any weight at all. We would be truly weightless, but our mass still would not have changed.

If we're sitting on a chair and we drop a book, the force of gravity causes the book to accelerate as it falls toward the Earth. In your hands the book was at rest and so had zero speed. When it hit the floor it had certainly moved, so it must have had a non-zero speed. This change in speed, or velocity, is acceleration. But this acceleration is not random. For example, the Earth's gravitational field causes the book, or anything else, to speed up from rest to a speed of about 32 feet per second after only one second. After two seconds, its speed would be about 64 feet per second, and so on. More precisely, on the surface of the Earth the acceleration due to gravity is about 32 ft/s^2. This value varies somewhat with latitude because the Earth isn't a perfect sphere. Just like the author of this book, the Earth is somewhat fatter in the middle than at the top and bottom. Gravity also varies with altitude. On the top of a mountain, which is farther from the center of the Earth, the gravity will be slightly less than at sea level. When we combine the slightly lower value of gravity with the decreasing air density, it is not surprising that the Olympic long-jump record which was set in Mexico City in 1968 (at an altitude of about 1.4 miles) remained unbroken for almost twenty-five years. Long-jumpers, discus throwers, archers, and other competitors who must factor air resistance into their sports find that they can do a better job at higher altitudes. Marathon runners and other endurance athletes, who must rapidly replenish their expended energy stores with oxygen from the environment, find it more difficult to compete at higher altitudes. Nevertheless, they often factor high-altitude training into their routine. If they can get used to running at a two-mile altitude, then return to near sea level to compete, their body will feel a big boost in energy from the relative increase in oxygen they now have at their beck and call.

Galileo was one of the first to try to understand the difference between mass and acceleration. According to popular legend, he

Figure 29. The *Apollo 15* astronauts dropped a hammer and a feather on the Moon to see which hit first. As Galileo predicted three hundred years earlier, both hit the surface at the same time. (Courtesy NASA)

dropped two balls of different masses from the Leaning Tower of Pisa and noticed that both were accelerated by the Earth's gravity at the same rate. The force on them was not the same, but their acceleration was. The *Apollo 15* astronauts tried this same experiment on the Moon. They dropped a hammer and a feather to see which would hit the ground first (fig. 29). As expected, in the vacuum of space both hit the ground at exactly the same time.

Atmospheric pressure is the weight of the atmosphere per unit area. The weight of the column of air on top of your head that starts at sea level and reaches all the way to space is about 400 pounds. A column of air that measures one inch by one inch would have a weight of about 14.7 pounds. Dividing this weight by the area gives the atmospheric pressure, which is 14.7 lbs/in^2 or 14.7 psi. In metric units, pressure is measured in N/m^2. This unit also has a special

Out of This World

name called the Pascal, named after the French scientist and philosopher Blaise Pascal, and is abbreviated simply as Pa. One thousand Pascals is one kiloPascal (just as one thousand meters is one kilometer), which is abbreviated kPa. The typical atmospheric pressure in metric units is about 100 kPa. These units, kPa or lb/in^2, are the units you will find on the gauge that you use to check your tire pressure. The gauge measures the pressure in the tire, relative to the outside air. Both the tire gauge and the barometer can be used to measure pressure. Because the barometer came first, it was easier to simply read the height of the column of Mercury off the tube and stop right there, rather than use the height to calculate the mass of the column and then use the mass to calculate weight and pressure. For this reason, your local weather forecaster often reports pressure in inches of mercury. We will allow them this luxury, but let's remember that pressure is really a force per unit area.

In space, there is almost no atmosphere. There is still some, though not nearly to the extent that we are used to having here on Earth. As we saw, at a 50-mile altitude the atmosphere contains far less air than it does on the surface. At 200 miles, where the International Space Station is located (fig. 30), the air is less than one one-billionth the density of the Earth's surface. That means you would have to take one billion gasps just to get a single lung full of air. For astronauts and cosmonauts, that's as good as zero. There is not nearly enough of an atmosphere in space to support life, so we refer to space as a vacuum. Aside from the obvious problems that working in a vacuum would pose for you and me (such as a sudden painful death as we die of the bends), a vacuum can also pose a problem for spacecraft. Without any atmosphere, there can't be any atmospheric pressure. Spacecraft, which are designed and assembled on the ground—under atmospheric pressure—will be subjected to a significant pressure change as they move into space where the pressure

Figure 30. The International Space Station will orbit at about a 200-mile altitude. At that height, the air is over one billion times less dense than on the surface of the Earth. (Courtesy NASA)

is essentially zero. If the spacecraft compartments are tightly sealed—to keep air inside for a crew to breathe, for example—then there will still be atmospheric pressure inside the spacecraft but essentially zero pressure outside of it. The air inside the crew compartment will literally try to explode the spacecraft, just as the carbon dioxide in your soda bottle tries to explode the bottle as you loosen the top. The spacecraft structures must be strong enough to resist this force. The wall of a crew compartment measuring 10 by 10 feet would feel a force of over 211,680 pounds due to the air inside. A spacecraft carrying a crew must be designed to contain this force, in the same way that a submarine must be designed to resist the implosive force from the water that surrounds it.

An interesting problem this poses for the designer is which way to open the door. To help keep the air inside, one would like to have the door open inward so that the force of the air helps to keep it closed. Similarly, one would think that submarines should have doors that open out, so that the water helps to keep them closed. However, the force of the air inside the spacecraft would be so great that a crew could never get the door open. As a result, spacecraft are designed to have hatches that open outward, (and submarines are designed to have hatches that open inward). It makes the job of sealing them more difficult, but it ensures that the crew can get them open when they need to.

It is possible to try and take shortcuts by using pure oxygen in the cabin, rather than the nitrogen/oxygen mixture we find on the ground. Recall that our atmosphere is really about 78 percent nitrogen. Since our bodies need only the oxygen to survive, if we eliminate the nitrogen we should be able to get by on an atmosphere of pure oxygen with much lower atmospheric pressure, right? This is absolutely true, and some past missions have done this. Unfortunately, it is also extremely dangerous. If a fire starts in a trash can on the Earth, you can run into the next room for a fire extinguisher, then run back and have a good chance of putting the fire out. In a pure oxygen environment the fire will burn so quickly that you won't have time to do anything but scream. This lesson was learned the hard way on *Apollo 1*. Initially, the spacecraft was designed to use a high-percentage oxygen atmosphere at a lower pressure. When a fire started during a ground test, it spread so quickly that the astronauts perished within a few seconds. The decision was made to use normal atmospheric pressure, and normal atmospheric composition, on future flights to make sure that this problem never happens again. As with every rule however, there are exceptions. Astronauts who must go outside in their space suits can't move if the suit is pressurized at the normal atmospheric pressure. If the glove you are wearing were crammed full of air at 14.7 psi, you would have to exert quite the force on it just to be able to close your fingers and pick something up. No matter how strong you are, you just can't do that in space at 14.7 psi. To make it possible for the astronauts to pick things up, their space suits are only pressurized at about 5 psi, using a high oxygen content atmosphere. Even then, working inside a space suit is extremely tiring because every move requires a fight against the pressure in the suit.

If we find a way to deal with all of the issues associated with atmospheric pressure, we must also bear in mind that the atmosphere helps to insulate us from hot and cold extremes. If you get too hot or too cold in the room you're in right now, you can probably adjust the temperature by turning on the air conditioner or heater, which blows either cooler or warmer air into the room. In space, there is no air, so the spacecraft can't use it to warm up or cool down. Without the atmosphere to insulate a spacecraft, surface temperatures can range from over +120° F during sunlight—hotter than Death Valley, California—to less than −70° F during sundown, which is colder than the average temperature at the South Pole. In addition to providing air to breathe, spacecraft and space suits also need to protect a crew against these temperature extremes.

Last but not least, the atmosphere also protects us from highly energetic radiation from space that strikes the Earth on a regular basis. Some of this radiation comes from the Sun and some comes from far outside our solar system. Although the Sun looks yellowish orange, because it emits mainly yellowish orange light, it actually emits light of many different colors. When combined together, our eyes interpret all of these colors to be white. When you are looking at a piece of white paper, you actually see light of all different colors being reflected from the paper to your eyes. When you print black

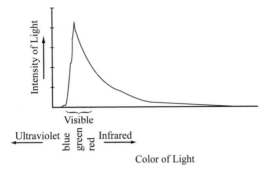

Figure 31. The Sun emits light of many different energies, or colors. Our eyes are sensitive to those energies where most of the light is generated. This is what we call visible light.

words on a white piece of paper, the words appear black because the ink does not reflect any colors of light well. In other words, you are not seeing any light reflected off the black print. When you draw on the white paper with a red crayon, the crayon markings appear as red because only the red color of light is reflected by the crayon. The Sun emits not only light that our eyes can see (red, green, blue, and so on), it also emits light that cannot be seen by the human eye (fig. 31). It is interesting to note that "visible" light is the spectrum of colors where the Sun emits most of its light. In other words, our eyes are designed to respond to the colors of light that the Sun generates. If our Sun were hotter, or cooler, it would not emit most of its energy in what we call the "visible" band. The band would have shifted to other colors of light, and our eyes would probably have adapted to view other colors as being visible. All colors of light, both visible and invisible, form what is called the *electromagnetic spectrum* or *electromagnetic radiation*. In deference to our physical limitations, we typically reserve the term "light" to refer to light which is visible, and we refer to the nonvisible as "electromagnetic radiation."

We usually don't think of nonvisible electromagnetic radiation as having color since our eyes can't see it. However, scientists tell us that color is simply one means of describing the amount of energy contained by the radiation. Violet light is more energetic than blue light, which is more energetic than green light, which is more energetic than red light, and so on. Light that is a little too low in energy to be seen by our eyes is called *infrared*. It lies "beneath the red" in terms of energy. The remote control device you use for your television or VCR probably works in the infrared. You can't see the infrared light emitted by the remote control, but it's there. If someone is standing between the remote and the television, the television won't respond because the infrared light can't go through the person's body. Light that is a little too high in energy to be seen is called *ultraviolet*, or simply UV. It lies "above the violet" in terms of energy. We can't see ultraviolet light with our eyes, but cats can. White shirts seem extremely bright when ultraviolet light is shined on them, as discos and soap-commercial companies well know. The higher-energy ultraviolet light emitted by the Sun is absorbed by the ozone layer in the Earth's atmosphere and can't reach the ground. This is fortunate for us because a single particle of ultraviolet light has enough energy to physically change many materials. If solar ultraviolet light were to strike the Earth's surface, it would damage many living things, but especially our skin. This is why increased solar exposure leads to increased chances of skin cancer and why millions of dollars are being spent on high-UV-protection sunscreens each year. On the Earth's surface we are protected as long as the ozone layer remains in place. In space, where there is no ozone layer to absorb the ultraviolet light, people—and spacecraft—can have problems.

Obviously, the atmosphere does many things for us, and its absence would pose some significant challenges. However, space is more than just the absence of an atmosphere. It is also important to bear in mind that everything is relative. On the surface of the Earth, the atmosphere contains over 10 billion billion molecules per cubic inch (that's a 1 with 19 zeros behind it). When we reach Low Earth Orbit (LEO) this number has fallen to less than one-billionth of this value, not nearly enough to keep you and me alive. However, one billionth of 10 billion billion is still 10 billion, a very large number. In LEO there is still enough of an atmosphere for a spacecraft to notice even if our human bodies can't. As we saw earlier, for a space-craft to remain in orbit it must be traveling incredibly fast—about 18,000 miles per hour. At this speed the spacecraft has enormous kinetic energy, which can give rise to two significant interactions with the atmosphere: sputtering and drag.

Sputtering is a term most of us are not familiar with. It is a technical expression that is most often used to describe the process used to place thin coatings onto some surfaces. For example, watch manu-facturers bombard steel watch coatings with highly energetic gold atoms. If done at the right energies, the gold atoms will stick to the surface. This is called sputtering. The same term is used to describe what happens when the incoming atoms strike with enough energy not to stick, but to break atoms loose from the surface—a possibility when materials are put into space. They may be sputtered, literally eroded away just like a sandblaster takes paint off a wall, by the environment. Fortunately, for most materials this does not appear to be a big problem in Earth orbit. Around other planets, it might be more of a concern because orbital velocity for that planet may need to be far greater than the 18,000 mph our spacecraft travels around the Earth.

Of much more concern for most spacecraft is the drag force. Drag is the reason your dog's tongue flaps in the breeze behind him when he sticks his head out the car window. It is the term used to describe the force of the atmosphere on an object that is trying to move through it. Simply stick your hand out the window of your car as you're driving down the street, and you'll feel the wind exert a force on you. That's the drag force. In space, drag from the atmosphere will cause a spacecraft to slow down, drop to a lower altitude, and eventually reenter the atmosphere and burn up. Designing spacecraft to counter the effects of the drag force is an important part of spacecraft design. After all, there's no point in spending millions of dollars to get a satellite into orbit if it doesn't stay in the orbit long enough to do anything useful.

In addition to being far thinner than that found near the ground, the atmosphere in LEO has a different chemical makeup as well. Near the Earth's surface we always find two nitrogen atoms, or two oxygen atoms, chemically bound together. Chemists would call this the "molecular form" of these elements. The molecular forms of nitrogen and oxygen are very stable, and they don't react very well with other chemicals. By the time we move up to LEO, we find a very different environment. The thin atmosphere is about 99 percent oxygen, and it is not molecular oxygen, but *atomic oxygen*. The solar ultraviolet has broken the molecular bond and caused the oxygen molecules to split apart into single atoms of oxygen. Because single oxygen atoms are lighter than either molecular nitrogen or molecular oxygen, they can move faster and get farther from the Earth's surface before gravity eventually slows them down. This is one reason why the chemical makeup of the atmosphere changes in LEO.

Atomic oxygen is very reactive chemically. When it hits spacecraft surfaces it can degrade materials, much like paint remover can take the paint off a wall. This is a big concern on longer missions, like the thirty-year plan for the International Space Station. A little bit of atomic oxygen could completely erode away many surface materials in a few years. This forces spacecraft builders to use expensive coatings on many materials to extend their useful lifetime. Atomic oxygen also gives rise to the well-known "Shuttle glow" seen near the surfaces of many materials in space (plate 6). Glow occurs when oxygen atoms bounce off the surface of a spacecraft with a little extra energy and then emit that energy as a visible light. Atomic oxygen is a constituent not only of Earth's atmosphere, it is also found on Mars and other planets of our solar system; so these interactions are a concern in other orbits as well. Interactions with the atmosphere could pose problems around any planet a spacecraft might orbit.

As we move away from LEO into higher orbits or beyond, most of the effects of the atmosphere become so small that they can be ignored completely. Unfortunately, when we leave these problems behind, others start to become more noticeable. Just as the chemical makeup of the atmosphere changed, the electrical nature of the atmosphere changes as well. On the ground, the atmosphere is essentially uncharged; that is, it is neither positive nor negative. Like mass, charge appears to be a fundamental property of matter. We can find evidence of charge easily enough. Just try sliding your feet over the carpet on a dry day. It is not unusual to feel a shock afterwards as we reach for a doorknob or shake hands with a friend. This shock is evidence of the electrical charges present in nature. Charges come in equal numbers; that is, positive charges always come balanced with equal numbers of negative charges so that the world as a whole is uncharged. The act of sliding your feet across the carpet, however, removes some of the negative charges from the carpet and you remain "charged" until the shock removes the excess charge buildup from your body.

These charges come from the subatomic particles that make up the atoms and molecules that form the world as we know it. An atom is built from subatomic particles called *neutrons*, *protons*, and *electrons*. For example, an oxygen atom is made up of eight protons and eight neutrons in the core, called the *nucleus*, which is surrounded by eight electrons (fig. 32).

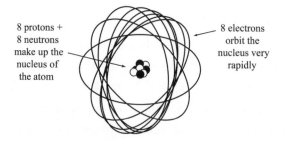

8 protons +
8 neutrons
make up the
nucleus of
the atom

8 electrons
orbit the
nucleus very
rapidly

Figure 32. A simplified view of the oxygen atom, consisting of eight protons and eight neutrons in the nucleus, plus eight electrons orbiting as shown. The electrons orbit a distance of about 10^{-8} inches, while the nucleus is about 10^{-13} inches in size.

Neutrons are electrically neutral. That is, they have no electrical charge at all, hence their name. Protons have a positive electrical charge, and electrons have an equal but opposite negative electrical charge. Eight positives and eight negatives balance out, so the oxygen atom is neither positive nor negative and is said to be electrically neutral. As a result, an oxygen atom wandering around on its own would not respond to electrical forces. But there's a problem. Our old friend the solar ultraviolet, which is very good at breaking chemical bonds, is also capable of removing an electron from many of the oxygen atoms. When this happens, an electron goes on its merry way separate from the oxygen atom, which is then left with eight positively charged protons but only seven negatively charged electrons. The resulting oxygen *ion* is left with a positive electrical charge. The atmosphere as a whole still contains equal numbers of positively charged oxygen ions and negatively charged electrons, but now some are physically separated from each other, so they can respond differently to electric or magnetic fields. Physicists call this kind of mixture of electrically charged particles a *plasma*. Back in the 1920s Irving Langmuir, one of the first plasma physicists, thought there was some similarity between the positive and negative charges wandering through his electrically charged gas mixtures and the red and white blood cells that wander through blood plasma, so he used the same term. But there the similarity stops.

Over two thousand years ago, the ancient Greeks thought that all matter was composed of four basic states: earth, water, air, and fire. Having learned much in recent centuries, we now know for a fact that there are indeed four basic states of matter, but we would call them: solid, liquid, gas, and plasma. Heating a solid, such as a cube of ice, would produce a liquid, water. Heating the liquid would produce gas—steam—and heating the gas (in this case by bombarding it with solar ultraviolet light) would eventually produce a plasma. As we move away from the Earth, the relative percentage of the atmosphere that is electrically charged increases from about 1 percent in LEO to almost 100 percent in high Earth orbit and beyond. We find some everyday plasmas inside the tubes for neon lights, and several years ago plasma spheres, which show lightning-like effects inside a small glass ball, came on the consumer market. The radio blackout that is associated with spacecraft reentering the Earth's atmosphere is the indirect result of a plasma cloud being formed by the heat of reentry (fig. 33). Farther out in our solar system, the Sun, for example, is a plasma. As a whole, over 99 percent of the Universe is in a plasma state. Our home, the third rock from the Sun, just happens to be a small piece of the 1 percent that is not.

The trouble with the plasma environment is that if the independently moving charged particles in the atmosphere strike a spacecraft in unequal numbers, they can charge the spacecraft to high electrical potentials. This concept, that a lack of balance can lead to problems, has many analogies in everyday life. Consider the problem of keeping a hot-air balloon on a stationary platform so that passengers can easily get on or off. Because the hot air balloon is designed to be lighter than air, there must be some way to keep it on the ground. We could do this by tossing a few bricks to the pilot. If

Figure 33. As illustrated in this artist's conception, the Space Shuttle generates a plasma cloud when it reenters the Earth's atmosphere. (Courtesy NASA)

everything were exactly balanced, the downward force on the balloon due to gravity would precisely balance the upward buoyant force of the balloon, which would remain stationary, or neutral. If we don't toss enough bricks to the pilot, the upward force dominates and the balloon takes off. If we toss too many bricks to the pilot, the downward force dominates. If the downward force gets too great, we may find that the weight of the balloon can collapse its landing platform. Just as a difference in gravitational forces can lead to problems, a difference in electrical forces can also lead to problems. If different surfaces on a spacecraft charge to different electrical potentials—and in the right plasma conditions they almost certainly would—an electrical collapse could occur in the form of arcing.

Arcing is something that most of us recognize when we see it, but no one seems to be sure exactly what it is. If we unplug a toaster from an electrical outlet without switching the toaster off we may see a little blue arc near the end of the plug as it pops free of the outlet. This is arcing in action. The problem with arcing is that if it happens in the wrong place it can lead to disastrous consequences.

One of my first experiences with arcing occurred when I was working on a control console in the laboratory one summer. My task was to re-create a schematic diagram of all the wiring connections because the originals had been misplaced or rendered useless by a series of modifications and add-ons. To do this, I had to pull the back off the panels to expose the wires and trace the wires from the power supply input to the specific control elements. My boss was very clear about the need to turn the power off at the breaker before I did this, and I religiously verified that the power was indeed off by checking each wire with a voltmeter before I touched it. After several weeks of flipping off the breaker and verifying that yes, indeed, the power was off, I flipped the breaker one day and found that one wire was still hot. I needed to check if there really was current flowing through the wire, so I leaned back to make sure I had a good view of the output. When I did, I managed to touch the metal chassis of the control panel with the metal probe I was using to monitor the current on the wire. This established an electrical connection from the wire to the chassis and caused a nice big arc inside the panel. All of a sudden I heard a loud "pow" and the inside of the panel was showered with blue sparks. I decided that getting out of the panel, without touching anything, would be a wise thing to do, but as I tried to leave, the metal probe didn't want to come with me. I looked up and realized that the arc had melted the probe onto the chassis. I was lucky to emerge undamaged, and I had to get a hammer to pound the probe loose before continuing.

There are big concerns with arcing. Not only do arcs kill, but an arc occurring where you don't want it to can melt wires or otherwise damage electrical circuits. Arcing on spacecraft is of big concern because, with rare exception, once a spacecraft breaks it is simply not possible to fix it with whatever equipment—or crew—are on board, so it is lost forever. As a result, engineers must typically take extreme measures to ensure that their spacecraft won't break.

Plasma storms are periodically encountered on the Enterprise of the future, but the futuristic electrical systems seem to be designed to "take a licking and keep on ticking." A good example of the steps that engineers will take to prevent arcing is seen even today on the International Space Station. The Space Station uses solar arrays to generate electrical power. Those arrays produce an electrical potential difference of 160 volts. (By comparison, most electrical outlets in the United States are 120 volts.) Just like dragging feet across a carpet, dragging the solar panels through a plasma causes an electrical voltage to build up. As originally designed, the Space Station structures could charge to as much as −140 volts. By itself, this would probably not be a concern because all of the Space Station structures would be at −140 volts. Engineers will tell you that, like a pressure difference, only a voltage difference can cause problems. So the Space Station will have a problem only if it comes in contact with something that is not charged at −140 volts.

Unfortunately, this would happen at any time that the Space Shuttle came up to dock. The Shuttle has a very different electrical power system and would have an electrical potential of at most a very few volts. The concern is, could a 140-volt potential difference between the Shuttle and Station lead to arcing or other problems? We believe the answer will be no because a voltage difference is only part of the problem. The other half of the problem is the electrical current.

In the United States, a 120-volt potential difference is considered the standard. At this voltage, most electrical outlets are capable of supplying 15 to 20 amperes, or amps, of current. The product of amps and voltage is the power, measured in watts. My toaster, for example, uses about 10.8 amps of current at 120 volts to consume 1300 watts of power. In LEO, the current available from the plasma environment is very low. The entire Space Station would draw less than 0.001 amps of current from the environment. As a result, a "worst case" arc would dissipate no more than 140 volts × 0.001 amps = 0.14 watts. This is a pretty low value, so the Space Station designers don't think that arcing will be a problem. But, as the sailing expression goes, they won't spoil a ship for a ha'porth of tar, so they're still going to take some precautions just to be safe. One of the first payloads carried into orbit for the Space Station will be a plasma contactor, an electrical device that is designed to maintain the potential difference between the Space Station and the environment at about −40 volts rather than the −140 volts it would otherwise be. Even if arcing could occur at that small voltage difference, the very low current should prevent any damage from occurring.

As if electrical charging weren't enough of a concern for space travel, another component of space called the *radiation environment* is also present. Earlier we used the term "electromagnetic radiation" to refer to "light" of different energies. Spacecraft builders also use the term "radiation" to refer to higher-energy particles that can penetrate through many materials and damage electrical components or the crew. Some of this radiation—gamma rays and X-rays—is part of the electromagnetic spectrum we discussed earlier. More usually though, the radiation environment is made up of particles such as neutrons, protons, or electrons. In LEO, the most plentiful source is the Earth's trapped radiation belts. The radiation is "trapped" around

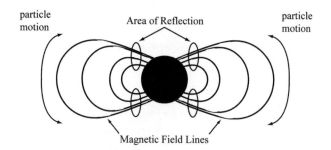

Figure 34. The Earth's magnetic field forces charged particles to bounce from north to south along the magnetic field lines. These charged particles are said to be "trapped" by the magnetic field and form the trapped radiation belts.

the Earth because charged particles must respond to electrical and magnetic forces, just as mass must respond to gravitational forces. Any charged particles produced near the Earth, including the plasma environment, feel the influence of the Earth's magnetic field. These charged particles literally become "trapped" by the Earth's magnetic field lines and wind up bouncing from north to south to north (fig. 34). These charged particles form the Earth's trapped radiation belts, which are sometimes called the Van Allen belts in honor of James Van Allen, the American physicist who made the first direct measurements of the radiation belts with an instrument on Explorer I, the first American spacecraft. Van Allen had put a Geiger counter on board and hoped to detect a few high-energy particles called cosmic rays. After launch, it seemed to him that the instrument didn't work because the data it was returning didn't make any sense. Then he realized that the instrument was working fine, it was just detecting about 10,000 times more particles than he had expected. This was the radiation belt. Planets with a larger magnetic field, like Jupiter, have even more intense radiation belts than Earth. Any planet with a smaller magnetic field, like Mars, will have weaker radiation belts.

The radiation belts deliver a dose of radiation to a spacecraft and its crew. Images of Hiroshima and Nagasaki, and accidents like

Table 6
Radiation Exposure on the Surface of Earth

Source	Dose Rate (Rad/yr)
Internal irradiation	
Potassium 40	0.020
Carbon 14	0.001
External irradiation	
Soil	0.043
Radon in air	0.001
Cosmic rays	
Sea level	0.035
1-mile altitude	0.04–0.06
2-mile altitude	0.08–0.12
3-mile altitude	0.16–0.24
4-mile altitude	0.30–0.45
Older technologies	
Radium watch dial (1960s)	0.04
Shoe X-ray fitting (1930s)	< 0.001

Note: Astronauts receive about 0.1 rad during a ten-day Shuttle mission.

Three Mile Island or Chernobyl, give most of us a fearful picture of radiation—and with good reason. Radiation will damage the molecular structure of most things and, if received in high enough doses, can cause big problems. On a spacecraft, the crew itself is the most sensitive to radiation damage; electrical components such as solar arrays, computers, and so on follow as a close second. All of us on Earth are continually exposed to a low dose of radiation from naturally occurring sources, as shown in table 6. In space (in LEO, to be more specific), astronauts can get ten times their yearly surface dose in just a few days. Fortunately, this appears to be a factor of one hundred or so below any safety thresholds, but radiation damage may ultimately limit the amount of time a crew can stay in orbit around a planet. In higher Earth orbits, or in orbits around other planets like Jupiter, the radiation belts are so intense that they would quickly be lethal to a crew unless protected behind several feet of shielding. Even on unmanned missions, spacecraft electronics will be exposed to the radiation environment and will, over time, be degraded. In many cases it is the cumulative effect of many years' exposure to radiation that causes today's spacecraft to fail.

As if the preceding problems weren't enough, another issue that spacecraft must deal with is the possibility of damage from micrometeoroids (see below) or orbital debris. In the past couple of years, Hollywood has produced a number of motion pictures illustrating the grave peril from comets of doom or asteroids the size of Texas. History has recorded quite clearly that these large bodies are indeed found in nature. On June 18, 1178, the occupants of the English monastery at Canterbury reported seeing "a flaming torch" emanating from the Moon which was "spewing out, over a considerable distance, fire, hot coals, and sparks." This catastrophic event was undoubtedly the result of a collision with an extremely large asteroid or meteoroid. A little closer to home, in 1908, about 1500 square miles of the Podkarrennaya Tunguska River basin in Russia was completely devastated by a large meteoroid that impacted with the energy of about 10 megatons of explosive. The large crater in the desert in Arizona, which measures 4150 feet in diameter and 575 feet deep, is evidence of another large impact (fig. 35). It has been proposed that the extinction of the dinosaurs was the indirect result of a large asteroid impact in what is now the Gulf of Mexico, with the smoke and dust generated by the impact causing the Sun to be obliterated and much vegetation and many animals to die. Fortunately, the probability of running into these doomsday asteroids, or of them running into us, is so remote that spacecraft designers tend to ignore them altogether. After all, if something the size of Texas were in your

Figure 35. This giant crater in the Arizona desert is thought to be the result of a large meteoroid impact thousands of years ago. (Courtesy NASA)

way, you would be able to see it and maneuver your spacecraft out of its path. It is actually the very tiny cousins of these monster asteroids that pose more urgent problems.

Micrometeoroids are small pieces of matter, usually the size of grains of sand or smaller, that are a natural part of our solar system. They are presumably the leftover pieces of larger comets or asteroids. On Earth we see evidence of these micrometeoroids anytime we see a shooting star. A shooting star is simply the glow created by a micrometeoroid as it burns up upon entering the Earth's atmosphere. (Another beneficial thing the atmosphere does for us is to protect us from micrometeoroids.) Most of these shooting stars are quite small and burn up quite rapidly. By the time you can say, "Hey! Look!" they are gone. Still, if the Space Shuttle, traveling at 18,000 mph, were to run into a one-inch micrometeoroid in space, it would cause the same damage that a 16-pound bowling ball would do to a car at 70 mph. Fortunately, as big as the Space Shuttle is (the payload bay alone is about 16 feet wide and 65 feet high), the probability of hitting a piece of matter big enough to penetrate it is very low. However, it is always a risk.

For the past forty years, we have made the problem worse by leaving behind in space little pieces of junk, called *orbital debris*. Orbital debris is a broad term used to describe anything that is man-made that gets into orbit, aside from the spacecraft itself. For example, throughout the 1960s and 1970s designers used explosive bolts to separate the nose cone from the launch vehicle. The cast-off pieces of bolts went into orbit, and they have stayed in orbit as orbital debris. (If you look closely you can even see them in the 1997 movie *Apollo 13* when the Service Module separates from the Saturn V Launch Vehicle.) Astronaut Ed White's first U.S. space walk in 1965 made the cover of *Life* magazine (fig. 36), but it also generated some orbital debris when his glove floated out of the open hatch of the capsule. Although a glove is very light, getting slapped on the face with one as the precursor to a duel still smarts. Slamming into one at the speed of 18,000 mph would release the energy of almost 100,000 face slappings. If you were receiving only one slap per day, it would take you 274 years to receive them all. If the slaps were delivered all at once, as would be the case if you ran into the glove on orbit, the result would be a rather unpleasant way to die. After about forty years of humans dirtying up space, there is just as much man-made debris in some orbits as there are naturally occurring micrometeoroids. A Pegasus rocket that broke up in 1996, for example, produced over seven hundred pieces of orbital debris alone. As spacecraft make the move to higher and higher speeds, the size of particle that is required to do significant damage becomes smaller and smaller. (Going at 99 percent the speed of light, a glove would impact with the energy of the atomic bomb dropped on Hiroshima.) This makes the risk of encountering debris go up dramatically. If we

Figure 36. Ed White's first U.S. spacewalk in 1965 made headlines around the world, but it also generated some orbital debris when a glove floated out the open cabin door. (Courtesy NASA)

don't stop producing this debris soon, we may need to rely on thick layers of shielding to protect future crews.

As we move away from the Earth into interplanetary space, we find that the relative abundance of the various environments changes. The neutral atmosphere near the Earth has become so thin that we can finally neglect it altogether. The few particles that are encountered in interplanetary space are all energetic enough to be considered a form of radiation. The Earth's trapped radiation belts are gone, but very energetic particles from the Sun are there. Periodically, the Sun is seen to hiccup and spew forth a very energetic blob of matter, consisting mainly of protons and electrons (plate 7). These occurrences are called *solar particle events*. Although infrequent, a single solar particle event can do more radiation damage to a spacecraft than a whole year's worth of exposure to the trapped radiation belts. Depending on the severity of the event, it can be catastrophic. During August 1972, a major such event occurred. Had an astronaut been on the surface of the Moon, shielded only by a space suit, the radiation dose could have been lethal. In orbit the Earth's magnetic fields help to shield us from solar particles, but they don't help at all once we reach interplanetary space. Moving away from the Sun decreases the dose, while moving toward the Sun increases it.

Finally, spacecraft are also exposed to high-energy particles coming from other star systems and not just our Sun. These particles are called *galactic cosmic rays*, the same ones that Van Allen was looking for with *Explorer 1*. They are perhaps evidence of stellar particle events on other star systems and add a small radiation dose to the spacecraft, but we ignore it at our peril. If we try to move through interstellar space at near light-speed velocities, the few particles we encounter would pass through our spacecraft with such enormous energies that they would deposit large doses of radiation on our subsystems and crew. Even if it were shielded behind several feet of lead, the radiation dose would probably be lethal to a crew within days.

In addition to the total radiation dose, which can cause permanent damage to electronics and the crew, both galactic cosmic rays and solar particle events can cause temporary problems as well. Many electrical devices are so small that the passage of a single energetic particle can cause them to malfunction—by corrupting data, for example (fig. 26). These *single event upsets* lead to another factor in the high cost of today's space electronics. All systems must be redundant so that the spacecraft can continue to function in a number of different ways if a single part or system malfunctions, or we must utilize specially made and very expensive parts that won't malfunction in the radiation environment of space.

Future spacecraft may also have to deal with weapons environments like phasers or photon torpedos, both considered a special case of radiation by today's definition. Radiation weapons are a real possibility. They may be as routine as setting off a nuclear warhead in space, or more sophisticated like the X-ray lasers that the U.S. Ballistic Missile Defense Organization (colloquially known as "Star Wars") program of the 1980s investigated. Regardless, radiation from an attacking Borg vessel could easily dwarf the naturally occurring radiation from all natural sources.

We can hypothesize about the environment found in interplanetary or intergalactic space, but since humans have never been there we are not quite sure yet what we will find. *Pioneer 10*, launched in 1972, and *Voyager 1*, launched in 1978, are now the most remote man-made objects from the Earth. They are over 6 billion miles away. *Pioneer 10* is on a path toward the constellation Taurus the Bull and should reach the next star in another 2 million years or so. Since no other spacecraft has gone farther from the Earth, no direct data on what lies beyond exist. We expect that the radiation environment stemming from galactic cosmic rays would remain fairly constant throughout interstellar space. When we approach other star systems, we may find the environment to be more harsh, or more benign, depending on the nature of the star in question relative to our Sun. Black holes or neutron stars would be bad news. Young, cool T-Tauri stars might be quite pleasant. In intergalactic space, the galactic cosmic ray environment should drop significantly, and intergalactic travelers would notice an environment that is as close to that of a complete and total vacuum as possible. However, that experience will remain well out of our reach for the foreseeable future.

With all the diverse environmental factors (radiation, plasma, micrometeoroids, and so on) that we have to contend with, space may appear to be a rather forbidding place. It is the task of the spacecraft designer and operator to make sure that the environment of space, and its numerous possible effects, goes unnoticed by the crew. As technology advances, this will surely become easier, just as it is becoming easier for automobile designers to make cars capable of traveling in various off-road conditions while still carrying their passengers in unprecedented levels of comfort. Surely, many things will continue to change in the coming years. Foremost among them may be the very reason for going into space, which is the subject of the next chapter.

To Boldly Go:
The Reasons for Space Exploration

To seek out new life and new civilizations . . .

The *Star Trek* mission is a bit open ended. We sometimes compare its attitude of "let's go see what's out there" to Christopher Columbus's voyage to discover a shorter passage to India. What is often overlooked is that Columbus was not motivated entirely by an academic desire to do something no one had ever done before; he was driven—at least in part—by capitalism. If he were able to find a shorter route to India, he could bring goods back from there to Europe in less time than it was taking everyone else. This would reduce the transportation costs, get the goods to market more quickly, and improve the profit margins of his investors. In other words, Christopher Columbus had a very clear mission.

Whether dealing with cars, ships, aircraft, or spacecraft, before the task of designing a vehicle can begin it is necessary to define explicitly the mission for the craft and its crew; that is, the planners have to answer the question, "What do we want it to do?" If they are looking for an inexpensive way to get themselves to and from work every day, and they eliminate the possibility of walking or bicycling, they will probably wind up looking at a small no-frills motor car. If

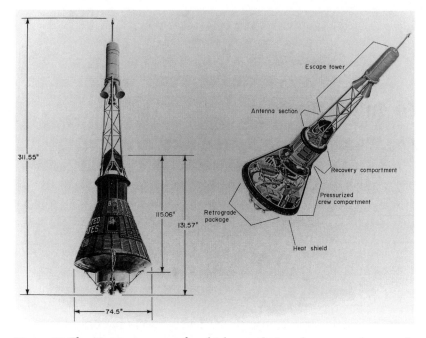

Figure 37. The Mercury spacecraft, which was designed to prove that people could function in space, was just big enough to hold one person sitting down. (Courtesy NASA)

their desire is to travel in style and comfort, they will examine a different variety of car, in a correspondingly different price range.

The first U.S. manned spacecraft program, Project Mercury (fig. 37), had simple objectives: to orbit a manned spacecraft around the Earth, to investigate man's ability to function in space, and to recover both crew and spacecraft safely. Project Gemini (fig. 38) had more ambitious goals: to subject crew and equipment to space flight up to two weeks in duration, to rendezvous and dock with orbiting vehicles, and to maneuver the docked combination by using the target vehicle's propulsion system. Finally, Apollo was designed to fulfill the mission "to put a man on the surface of the Moon and return him safely to Earth." To accomplish this, the Apollo spacecraft was very different from its predecessors (fig. 39). Not only was it larger,

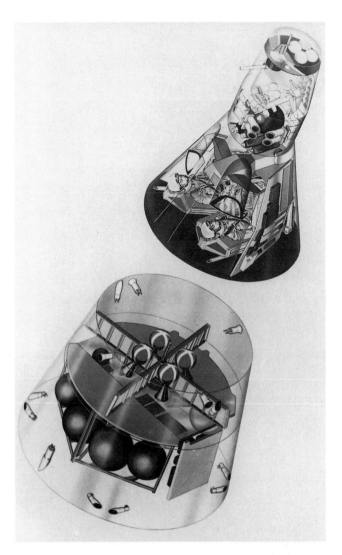

Figure 38. The Gemini spacecraft, which was designed to support a two-week mission in space, added a second chair, and also a second person to occupy it. (Courtesy NASA)

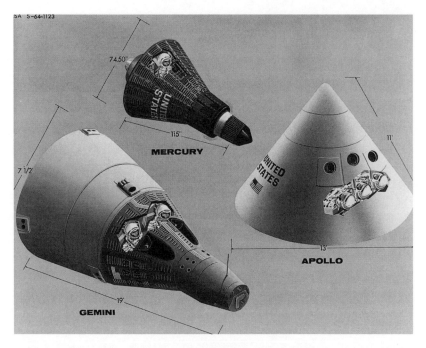

Figure 39. The Apollo spacecraft, designed to go to the Moon and back, added a third chair, a third person, and enough room to get in and out of the Lunar Module. It is shown here compared to the Gemini and Mercury spacecrafts. (Courtesy NASA)

to accommodate a crew of three, but it had a Lunar Module (called the LEM)—to enable a landing on and blast off from the lunar surface—and a much more robust Command Module and Service Module. Similarly, the Space Shuttle looks vastly different from any of *its* predecessors because one of its mission objectives was to be reusable (fig. 40). This meant that it had to survive reentry (as we have already discussed) in good enough shape to fly again.

A spacecraft is usually just a means to some end—a piece of machinery that will enable something to happen. For manned missions, it has to keep the crew alive and to get them from the ground to their final destination—an orbiting space station or a remote planet, for example. For unmanned satellites, the mission may be far differ-

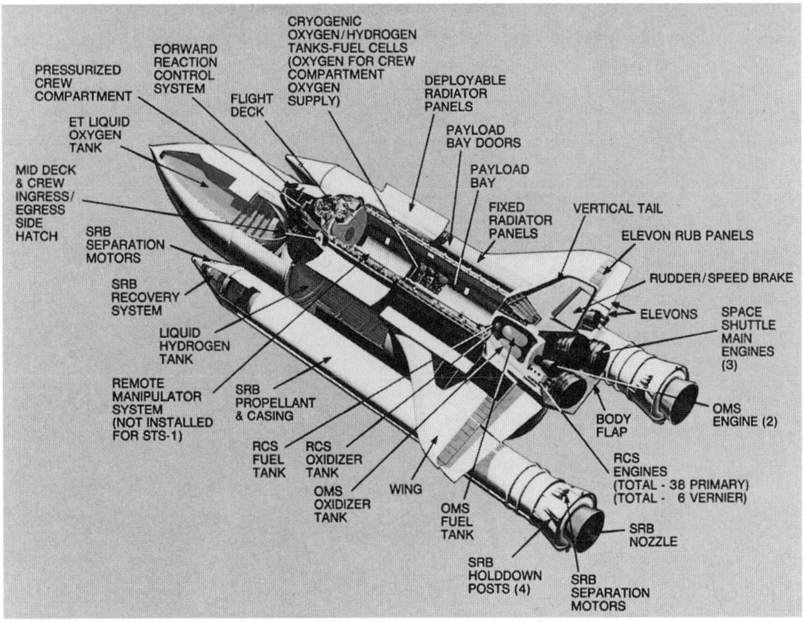
CRYOGENIC
OXYGEN/HYDROGEN
TANKS-FUEL CELLS
(OXYGEN FOR CREW
COMPARTMENT
OXYGEN
SUPPLY)
FORWARD
REACTION
CONTROL
SYSTEM
PRESSURIZED
CREW
COMPARTMENT
FLIGHT
DECK
DEPLOYABLE
RADIATOR
PANELS
ET LIQUID
OXYGEN
TANK
PAYLOAD
BAY DOORS
MID DECK
& CREW
INGRESS/
EGRESS
SIDE
HATCH
PAYLOAD
BAY
FIXED
RADIATOR
PANELS
VERTICAL TAIL
SRB
SEPARATION
MOTORS
ELEVON RUB PANELS
RUDDER/SPEED BRAKE
SRB
RECOVERY
SYSTEM
ELEVONS
SPACE
SHUTTLE
MAIN
ENGINES
(3)
LIQUID
HYDROGEN
TANK
REMOTE
MANIPULATOR
SYSTEM
(NOT INSTALLED
FOR STS-1)
SRB
PROPELLANT
& CASING
OMS
ENGINE (2)
BODY
FLAP
RCS
FUEL
TANK
RCS
OXIDIZER
TANK
RCS
ENGINES
(TOTAL - 38 PRIMARY)
(TOTAL - 6 VERNIER)
OMS
OXIDIZER
TANK
WING
OMS
FUEL
TANK
SRB
NOZZLE
SRB
HOLDDOWN
POSTS (4)
SRB
SEPARATION
MOTORS

Figure 40. The Space Shuttle is much larger, and much different in shape, than any of its predecessors because it had a very different requirement—to be reusable. (Courtesy NASA)

ent but will still require significant planning on the part of the designers. Obviously, one of the uses of space is, at least partly, political. The space race of the 1960s was clearly a race between nations to see who could perfect the necessary technology more quickly. With these issues of national pride aside, a more cost-justifiable means is usually needed, especially if one wishes to attract financial backing from private industry.

Watching the latest *Star Trek* epic from the comfort of a movie theater will convince the audience that the Starship *Enterprise* comes complete with a suite of sensors capable of locating a single individual on a planet inhabited by billions of people within a matter of seconds. This capability, high-resolution scanning from a great distance, is complemented by the medical tricorders that are capable of near instant diagnosis and cure. Regardless of the distances involved, these devices are capable of sensing vast amounts of data from a remote location. This is the whole premise behind *remote sensing*—literally, observing something from a distance.

Even today, remote sensing is a commercially viable use of space that is routinely exploited. Just as today's X-ray machines and Magnetic Resonance Imaging (MRI) devices may be the primitive predecessors of tomorrow's tricorders, today's spaceborne remote-sensing platforms surely have a great deal in common with the next generation. There are two distinct uses of space as a remote sensing platform. One is to observe the Earth from space. Weather satellites or spy satellites fall into this category, as do science missions designed to better understand the functioning of the global ecosystem. Another use is to observe space in ways not possible from the ground. A good example of this application is the Hubble Space Telescope (fig. 41). Many people believe that the primary advantage of placing a telescope in space is that it gets you much closer to the objects you want to observe, enabling you to get much sharper images than are possible with any ground-based telescope. Actually, this is not the reason at all. As we saw earlier, the closest star is about 6 million million miles away from the Earth, so being 50 miles closer would make little or no difference. In fact, placing a telescope in orbit in an attempt to get it closer to a star is the equivalent of moving about 100 molecular diameters closer to New York in order to get a better photograph of it from Los Angeles. Sure, you are closer to your final destination, but not enough to matter.

The advantage of the Hubble Space Telescope is that in space we are above the Earth's atmosphere. This provides us with three distinct benefits: (1) it is never cloudy, so we can observe the stars whenever we want to; (2) there is no atmosphere, so there is no turbulence to distort the picture; and (3) we are able to gather in all

Figure 41. Many people think that space telescopes, like the Hubble Space Telescope shown here, take sharper pictures because they are closer to the stars. The real reason space telescopes work better is that they are above the atmosphere so that the light from distant stars is not distorted or absorbed before it gets collected. (Courtesy NASA)

of the light reaching the Earth from the distant object. It is actually the Earth's atmosphere, and not the Earth's surface, that absorbs much of the light from the Sun. Any light reaching the Earth from straight overhead must travel through the 50-mile atmosphere to reach us. If we are looking closer to the horizon, the path through the atmosphere is longer still. Sunsets often take on a beautiful appearance because the light from the Sun must scatter off the thicker atmosphere—and the pollutants in it—making it a glorious deep red. Overhead, the light has to travel through far less of the atmosphere to reach us, and it appears yellow-orange. It is not generally appreciated, but the Earth's atmosphere lets only about 25 percent of the light from the Sun reach the surface directly. The remaining 75 percent is either scattered, absorbed, and used to heat the atmosphere, or reflected back to space. When we are above the atmosphere, there is nothing to get in our way, and that is why space telescopes are able to return much better pictures (plate 8).

In addition to taking visible-light pictures from space, it is also possible to take pictures of radiation not visible from the ground. Recall that the Sun generates light of many different colors, not only visible light. Other stars may generate light mainly in the infrared or ultraviolet spectrum, so by observing in different colors we can get much more data than is possible with visible light alone. This is why the Hubble Space Telescope, a visible-light telescope, is complemented with other space-borne telescopes such as the Extreme Ultraviolet Explorer and the Infrared Space Observatory. These telescopes are designed to look for light that is not visible from the ground. The Earth's atmosphere absorbs essentially all of the infrared and ultraviolet light coming to us from space. If we tried to look for these colors from the ground, even on a clear day we would not be able to see anything. From space we can see anything that's there. Data from space-borne telescopes provide astronomers with a more comprehensive picture of the Universe around us by giving us much more data than is available from Earth-based observatories alone.

Turning the cameras earthward opens a realm of other possibilities. It is possible to watch the shrinking of the rain forests, monitor the quality of the coffee crop in Colombia, and predict weather patterns from space. Accurate weather prediction is almost taken for granted today, and most television viewers can see satellite pictures of their local weather patterns every evening on the daily news. However, such predictions were not available until satellites were able to beam pictures back for the past thirty years or so. One factor in choosing the date for the Allied invasion of Europe on June 6, 1944, was the weather pattern. The Allied weather stations, which were located to the west, over the Atlantic, indicated clearing weather was on the way. The Axis weather stations, across Europe, were still reporting bad weather. The Allies knew that good weather was on the way, and that they could expect decent weather for a landing on June 6. This gave them additional hope that they could catch the Axis powers off guard.

In addition to getting visible images of our planet, which are very useful for weather forecasting, we can also get additional information through observations of specific colors. For example, sand or rocks will reflect brown light very well, while healthy vegetation reflects green light. Space telescopes often take pictures of the Earth at different colors, often by simply putting a filter in front of the lens, as you might do yourself with your own camera. By comparing observations of the Earth made at different colors, it is possible to deduce a great deal about the nature of the land being viewed. Grasslands, pinewoods, red sand, and silty water all will appear differently when viewed at different colors (plate 9). Learning about the Earth from space can help predict the quality of next year's crops, giving advance notice of any impending shortages in commodities. Imaging the Earth in nonvisible colors can also prove insightful. For example, on a recent Shuttle mission, a Spaceborne Imaging Radar device was used to search for evidence of China's lost Silk Road in the Taklamakan Desert (plate 10).

The Silk Road is the most well known ancient trading route of Chinese civilization. Beginning around 200 B.C., the trade route eventually grew to connect China with the Roman Empire. Genghis Kahn took the Silk Road west when he began his quest to conquer Asia; Marco Polo followed it east when he visited China in the 1270s. The route was used for almost 2000 years, but eventually it was replaced by sea travel, which proved to be quicker and more profitable. The new space images can help locate cities that have been lost for centuries beneath the desert sands, helping to guide archeologists

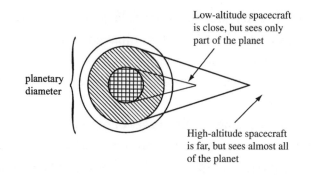

Low-altitude spacecraft is close, but sees only part of the planet

planetary diameter

High-altitude spacecraft is far, but sees almost all of the planet

Figure 42. Spacecraft in higher orbits can see more of the planet, but they are farther away so the images on the Earth look smaller. Spacecraft that are closer see larger images, but they see less of the planet at any instant in time.

to sites that will yield a wealth of data about prior civilizations. Other images from space indicate that China's Great Wall, which is the only man-made object visible from space with the naked eye, had extended far more than we knew. Finally, a radar imager was used by the Magellan spacecraft to map the surface of Venus, a task not doable in visible light because of the permanent, thick cloud cover that encircles the planet (plate 11).

Viewing pieces of our planet with radar or other remote sensing devices is certainly possible from the ground. However, ground observations are severely limited in the amount of surface they can see. Standing on the ground, one usually cannot see farther than a few miles from any particular spot on the Earth. The Earth has a total surface area of about 50 million square miles, so from any particular spot on the surface one can't see much of the whole. By going up in altitude, it is possible to see more and more (fig. 42). If you stand at sea level with your eyes about 6 feet above the ground, the horizon will be about 3 miles away. If you would like to see what it looks like 50 miles to the west, you will need to scale a 2100-foot tower. From 50 miles up in space, you can see over 500 miles to the horizon. Though it is more costly to get to space, it is easier and quicker to view different parts of the Earth from there, by simply adjusting the direction in which your telescope is pointing once you get there.

To appreciate how remote sensing actually works, we need to understand a little more about what is physically taking place. The whole premise behind remote sensing is that you deduce information about a distant object based on the information you receive from that object. When we see the Sun, for example, we are not looking at physical pieces of it; we are only seeing light that was emitted by the Sun. When you see a tree in the park, pieces of the tree aren't entering your eyes and being deciphered by your brain, even though some ancient Greeks believed this to be the case. You are seeing light—from the Sun—which reflected off the tree and into your eyes. When we "see" anything around us, we're really seeing the light that was either emitted by the object (the Sun, a lightbulb, a fire, a computer terminal) or light that was reflected off the object (trees, grass, the walls of your room). When designing a telescope or a camera to look at something, we have to have a pretty good idea about the kind of light that will come from the object in order to make sure that the device can actually see the object in question.

If we want to build a camera to take pictures of things around us, the camera had better be sensitive to visible light. But a camera can do more. It can be designed to collect light of any color, not just the visible spectrum. Although this type of camera would look very different from the usual camera, we can design it to look for electromagnetic radiation that is very different from visible radiation such as that used for radio, television, microwave ovens, or in X-rays. The signals emitted by X-ray stars or radio sources are physically very similar to visible light, except we cannot see them directly. A good example is the Crab Nebula. One of the first large radio telescopes was built back in the 1930s in an attempt to better understand the nature of the radio universe around us. The receiver showed a big burst of activity when the Sun was overhead, which was not surprising. However, it recorded other bursts of activity when the Sun was on the other side of the Earth. Eventually, people realized that one of the largest sources was actually the Crab Nebula, which was formed when a star became a supernova and exploded about 950 years ago. We know the date because of the numerous reports of a bright star in the heavens, in the exact location of the Crab Nebula, in 1054, lasting for over a year before it faded. This bright star was visible evidence that the star was becoming a nova. Hundreds of years later, we can watch how the ejected matter has been spewed off in visible light (plate 12), or we can even listen to it on the radio.

As mentioned in the previous chapter, if our Sun were hotter or cooler, visible light as we know it would be different. As blacksmiths have known for centuries, if you heat up a piece of metal it will eventually start to glow. At first, even though the iron is hot, it won't be visibly glowing, but if you get your hand close to it you can feel the heat rising. That's because the hot metal is emitting heat energy, in the infrared part of the spectrum. If the iron gets hotter, it will eventually start to glow in a dull red color. Get it hotter still, and the tip will start to become whitish. Heat it up even more, and it will start to emit in the ultraviolet. This principle, embodied in Planck's law of radiation, is that as an object gets hotter, the color at which it emits the majority of its energy changes. (The general shape of the light output is as shown for the Sun in figure 31. It is interesting to note that while the shape of the curve was known and agreed upon for years, physicists could not explain why it had this shape until the development of quantum mechanics.) We can apply this principle to camera design in two ways. First, we can use it to determine the temperature of distant objects by merely observing the color of the light they emit. We know that the surface of the Sun has a temperature of about 10,000° F not because we've run up close and stuck a thermometer into it, but because the light that is emitted by the Sun looks like it came from an object with a temperature of 10,000° F. Second, if we know the temperature of the object we are looking for, we know what color we need to look at to have the best chance of seeing it. This can tell us to what color of light our camera needs to be sensitive. For example, Americans, on average, have a body temperature of about 98.6° F. The British, for some reason, average about 98.5° F. Either way, because we have a temperature we are also emitting radiation, in the infrared. We can design a camera to look for the infrared radiation emitted by people. Some science centers and museums have such cameras so we can see who is "hot headed," but there is a more serious application: this is the operating principle of many night-vision devices. If you are flying overhead in an airplane trying to spot a downed aircraft in the California Sierras at night, you would have to hope that one of the crash survivors has a flashlight or give up until morning. Infrared devices can locate people by the body heat given off by survivors, or they can detect an aircraft's metallic surface, which retains heat better than the vegetation surrounding it.

Making a camera capable of seeing infrared light is a bit of a challenge, but it is getting easier with advanced technology. Today's designers use electrical-optics, or electro-optical devices. Basically, these cameras are designed to work in much the same way that our eyes work. Light received by an eye strikes the optic nerve, which converts it into an electrical signal that is then passed to the brain for deciphering. In the good old days, satellites used the same kind of film as cameras. In some cases, the film would be processed on board the spacecraft, then "faxed" back to the ground. On other missions, the film would be jettisoned from the satellite, reenter the Earth's atmosphere, and collected and sent to the lab for processing.

Today, electro-optical instrumentation, known popularly as digital cameras, can convert the signals directly into electrical currents, which can then be sent via radio broadcast back to the ground where they are reassembled. It may sound like rocket science, but your TV works on the same principle. The television station takes a picture, converts it into electrical signals, broadcasts them into the airwaves or transmits them over cable lines, and your TV reassembles the electrical signals into audio and video. It's no different with satellites, except the distances involved are far larger.

Another advantage that cameras, either digital or the film type, have over our eyes is that they can see fainter objects. One way they are able to do this is by collecting light for long periods of time. Our eyes operate continuously; that is, a picture in front of our eyes is continually updated and refreshed by our brain thousands of times each second. If we are sitting in a dark room, one without any visible light at all, we cannot stare at the wall in the hope of eventually gathering enough light to see it. As far as our eyes are concerned, if the light isn't there to be seen at this instant, we can't see it at all. With a camera, it's a different story. A camera can continue to gather light for long periods of time, allowing us to see fainter objects. This is the reason astronomers use time-exposure photographs.

As every amateur astronomer knows, another way to see fainter objects is to use a telescope with a larger lens or mirror. Light that enters our eyes has to come in through our pupils, which have a diameter of about half an inch. A telescope with a collecting area of one square yard would collect about 10,000 times more light than we could get with our eyes. This is another reason why telescopes offer a big advantage in remote sensing: not only can they be hooked up to cameras that can integrate the light emitted by dim objects for long periods of time, they can also collect light over a much greater

Figure 43. The Moon as seen normally (*left*) and under magnification (*right*). (Courtesy Celestron International)

area than our eyes. This means that much more light can be collected, so very dim objects can be seen.

In addition to gathering more light and making it easier to detect faint stars, telescopes can also magnify images. Our eyes come with only one power of magnification: one. We look at something and see it exactly as it is. If we examine the same object with a pair of binoculars or a telescope, the image is magnified, and we are able to see things that would otherwise be too small. When we look through a telescope, we are actually seeing an image of the object that is projected by the telescope. This image appears to be larger than the object itself (fig. 43).

All of the examples we have discussed so far are what we would call *passive* monitoring, that is, the telescope just sits back and takes what it can get from the source. If we try to take a picture with our camera and we don't have enough light, we can still take good pictures if we use a flash. The flash provides light that can reflect off our subject and back to the camera. This is an example of *active* monitoring. Instead of sitting back and waiting for light from the object to show itself, we create our own light and deliberately cause it to be reflected off the object that we want to see. Besides flashbulbs, another popular active technique is the use of radar, which was developed in the Second World War to help detect enemy planes or ships. Radar is an acronym that stands for *radio detection and ranging*. With radar, we simply create a source of energy—the radar beam—and send it out in the direction of whatever it is we want to see. The radar beam then bounces off its intended target, and some portion of that reflected beam then returns to our sensor. This is how the radar observations of Earth and Venus were made. Submarines use sonar, which is the same idea but with sound waves rather than radio waves. Knowing the time that passed between when the beam left and returned, along with knowledge of the speed at which the beam travels, tells us how far away the object is. This principle works the same whether you are listening for the echo of your voice off a canyon wall or are listening for the electromagnetic echo from an airplane. Radar plays a big role in making air travel safe by keeping track of all the aircraft in an area so they can avoid one another. Radar can also help pilots avoid obstacles in their flight path. We can, of course, use this active technique more broadly than just for radar or sonar. Radar uses radio waves, sonar uses sound waves; but there is also lidar (*light ranging and detection*), which uses light waves or laser beams. Both work on the same principle, but a laser beam is much more focused than a radio wave. Lidar can concentrate much more energy on a target, which also means we can see the target from farther away.

Let's turn to another application of remote sensing that has often been seen in *Star Trek*: scanning people with hand-held devices to assist in medical diagnosis. This idea is not as far-fetched as it might seem. X-rays have been used for over one hundred years to peer inside the body. More recently, it has become common practice to scan the interior of the body via Magnetic Resonance Imaging. MRI, which, unlike X-rays, does not deliver a radiation dose to the subject, looks for tiny magnetic signals from inside the body that are then interpreted by diagnostic computers. The *Star Trek* tricorder could be nothing more complicated than a tiny MRI device, coupled with a very complex software program that is capable of instant analysis and diagnosis. It may very well be that software capable of such rapid and complex analysis will be harder to develop than the instrument that gathers the data.

As is true of many commercial devices, there are also military uses of space. Spy satellites routinely take pictures of other countries' military installations to monitor troop movements or verify treaty compliance. According to rumor, such spy satellites—often in stationary orbits—can discern a newspaper headline from a height of 22,000 miles! We can only wonder how soon states or cities will have crime detection and prevention satellites of their own. Surely crime would decrease, or at least change noticeably, if remote-sensing satellites could follow the bad guys from the door of the bank they just robbed to their hideout and arrange for a warm reception by waiting police officers. We can get a taste of this today from helicopter news crews that watch highway police chases and broadcast them in real-time to the home viewing audience, as many will remember from the O. J. Simpson incident a few years ago. A popular rumor at the time was that prosecutors were trying to purchase Russian spy satellite images of Los Angeles in an attempt to pinpoint the location of O. J.'s white Bronco at the time of the crime.

Of course, knowing that they may be observed will motivate people to take steps to avoid being seen. There are a number of ways of doing this. One technique is to jam the radar source by simply sending back a larger radio signal so that the searching radar is overwhelmed—the equivalent of shining a floodlight into the lens of a passing photographer when you see the flash from her camera. Another technique is to create false targets to jam the detector. During the World War II invasion of France, many physicists spent the day bobbing up and down in small boats off the coast, tossing small strips of metal foil into the air. It was hoped that the reflections from the small strips of foil would fool German radar into thinking that a large number of Allied planes and ships were headed for parts of the coast.

In many cases, jamming is not even necessary because an incoming radar signal can itself be an early warning signal to the very people we are trying to find. If we send out a radar beam looking for the bad guys, the bad guys can also see our radar signal and tell where we are. The important part is that *they* can tell where *we* are long before *we* can find out where *they* are, as the makers of the popular "fuzzboxes" that detect police radar proudly proclaim. If we double the distance to the target, the intensity of the signal that reflects back to the source will decrease by a factor of four. In other words, if the police can detect that I'm speeding from a distance of 100 yards, I should be able to detect their signal from a distance of 200 yards. This gives me 100 yards to slow down to a legal speed. For this reason, many states outlaw radar-detecting devices for automobiles.

To make things even more complicated, if the bad guys don't want us to see them, they can design their equipment not to reflect radar, but to absorb it, much like the stealth fighter and stealth bomber. This stealth approach is not quite as effective as a Romulan cloaking device, but modern aircraft are designed to have small radar cross sections. Sharp curves and the all-metal exterior reflect radar very efficiently. The smooth lines and composite materials on the exterior of an airplane are designed to minimize reflection.

As time passes, remote sensing applications will continue to evolve, but they will still be based on the same fundamental principles discussed here. Telescopes may be larger, to gather in more

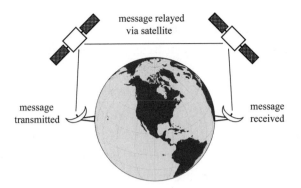

Figure 44. Satellites can be used to transmit messages from a ground station on one side of the world to a station on the other side.

light; digital cameras may become more sensitive, to respond to less energetic signals; and computer technology may be capable of finding the proverbial needle in the haystack. Yet the underlying physical principles will remain unchanged. Remote sensing is a cost-effective way of gathering data on a global scale, and the technique is expected to grow in significance in the future because of its wide possibilities for scientific, military, and commercial use.

Communications afford another prime example of a commercially viable application of space. The first transmission of live television across the Atlantic Ocean was provided by the *Telstar I* satellite, which was launched on July 10, 1962, and more advanced generations of Telstar communications satellites continue to be launched even today. If you want to make a telephone call to a friend on the other side of the Atlantic, or watch live television broadcast from the other side of the world, you have two basic choices. The signal can either be transmitted across a series of electrical cables that run between the two points, or the signal can be transmitted from the ground to a satellite and back, as shown in figure 44. Transatlantic cables have been in use for over 140 years and are certainly a very sure means of establishing the link, but they are both expensive and unforgiving of even a single point of failure. The first transatlantic telegraph cable placed between Ireland and Nova Scotia worked for only three weeks before failing.

Cables are expensive to use because we have to have enough of them to run from every point on Earth from where we just might want to send a signal, to each point where we might want to receive one. This might not be a big deal if we want to run phone lines in a crowded city where there are millions of people within a few 100 miles of one another. But what about the people living in rural areas? We have to run cables to their areas as well, or they will remain isolated. The cost of running electrical cabling all over the world is very high. Not only do we have to pay to mine the ore used in building the cables, but we have to manufacture them, and then pay somebody to install and maintain them. If we use 30 feet of copper wire per person (a reasonable assumption, since cables have to run into every house or office building that needs service), then we will need 30 billion feet of cable just to reach 20 percent of the world's population. For a small 20-gauge wire, this amounts to 400,000 cubic feet of copper weighing almost 50,000 tons. Given the current cost of copper (about $15 per pound), the cost of purchasing just the raw material would be about $1.5 billion. By the time we factor in the cost of converting the raw material into wiring, and installing the wires, it is easy to see that building the communications infrastructure of even a small country can easily become unaffordable unless it is spaced out over decades or centuries.

Newer technologies, such as optical fiber cables and microwave transmissions, offer a lower cost in comparison to copper cables. But the big advantage of satellite communications is that it offers lower total cost *and* easier access from almost any point on the globe. This is clearly evident by the explosion in the number of cell phones in use in recent years. A single ground station can "upload" signals to a satellite from hundreds of thousands of callers in a single cell (region) simultaneously. The satellite can then either rebroadcast them to callers in the same cell, or transmit the signals to other satellites on the other side of the Earth, which then beam them down. Although communications satellites are expensive to build and operate, they are more cost effective than the alternative of running electrical cables all over the place. If your friendly neighbor accidentally cuts your phone line while digging in his garden, you're out of business until the phone company can come out and install a new line. If a satellite goes down for some reason, the communications company can simply reroute your call through another satellite—and keep you in the loop while they do so. For this reason, many communications satellites are currently in use and more are planned. Motorola's Iridium constellation offers global phone service through its orbiting fleet of sixty-six satellites. The Iridium constellation (which circles the Earth in a 485-mile-altitude orbit) cost an estimated $5 billion to complete. Obviously, the undertaking would not have been funded by private industry if the revenue-generating potential were not significantly greater than the investment. Unfortunately, Iridium filed for bankruptcy in 1999 because many users could not afford the cost: about three thousand dollars to buy the satellite phone, plus four to five dollars per minute to use it. Undeterred, developers are planning an even larger fleet of satellites, called Teledesic, at a predicted cost of nine billion dollars. Teledesic would enable not only telephone communications, but also data transmission via satellite—a true Internet in the heavens.

Satellite television is another recent advance made possible by space travel. Individuals can purchase small satellite dishes, less than a yard in diameter, and obtain access to hundreds of television programs. When compared to the cost of running electrical cables around the world (the cable TV option) or broadcasting the signals of one hundred TV stations simultaneously to every region of the country, satellites again emerge as a very cost effective solution. To make this concept work, a user simply has to acquire a small dish and make sure that the dish is aimed at the proper satellite. Because the transmitting satellite is in geosynchronous orbit, its position, relative to the point of reception, won't change. Once locked onto the satellite, the user simply sits back and relaxes while his or her favorite shows are beamed directly down from space.

Relatively speaking, navigating a ship on the surface of the ocean or guiding a group of hikers on a wilderness retreat is pretty much a two-dimensional problem. We have to worry about north/south and east/west, but the complicating problem of up and down is simplified because we are captives on the surface of the Earth. But when we navigate aircraft or spacecraft, the up/down problem obviously becomes significant. One burning question is, "What's our altitude?" Another is, "What direction is down?" This may seem simple, but think again. In an aircraft on the Earth, if we drop a ball, gravity will pull it down. But as we have seen, weightlessness is a consequence of being in space, so a dropped ball doesn't fall, and a tumbling spacecraft can't easily figure out which way down is. Finding your way around in space is a tough problem.

As with communications, groups of satellites combined in what are called "constellations" can be used to help with navigation. These navigational constellations assist not only with space navigation but also with navigation on the surface of the Earth or in the Earth's atmosphere. The Global Positioning System (GPS) constellation of twenty-four satellites was funded by the U.S. Air Force as a means of navigating armed forces personnel and equipment (fig. 45). Each satellite broadcasts radio signals to users worldwide. By comparing the signals from different satellites simultaneously, it is possible to "triangulate" and compute your exact position to within a few yards'

Figure 45. The Global Positioning System (GPS) spacecraft.

accuracy anywhere on the surface of the Earth. Adding ground-based transmitters to augment the satellite transmissions can improve the accuracy to a few inches. Triangulation works by figuring out your relative distance from three or more well-known points. The GPS receivers work by broadcasting a timing signal from each satellite. The receiver deciphers the signals and computes the time difference between signals received from various satellites. Knowing the speed at which the signals travel, we can convert the time difference into a position difference. For example, your receiver may tell you that you are 1000 miles farther away from satellite A than you are from satellite B. Since we know the position of the satellites, by the time we add in our position relative to a third, fourth, and even fifth satellite from the constellation of twenty-four, we can determine our exact location to within a few feet. Although originally intended for the U.S. military (the Russian military has its own version, called Glonass), the GPS signals are available for use by anyone. Receivers can be purchased for a few hundred dollars, and prices are expected to fall in the future. These receivers enable trucking companies to keep accurate track of their fleet of vehicles and can help lost hikers find their way home. When U.S. pilot Scott O'Grady was shot down over Bosnia in 1995, he managed to establish radio contact with his comrades and was able to give them his exact location because he carried along a GPS receiver. Within minutes of learning his position, a rescue mission was dispatched, and the rescue was successfully completed within hours.

Even now, when you rent a car in some cities, you may find that it comes equipped with a GPS receiver. By coupling the receiver, which can determine your exact location, to a computer map of the city, the car itself can give you directions to your final destination. Because the GPS signals are updated continually, the car can alert you in advance to "take the next exit." If you make a wrong turn, the computer can immediately replot your travel plans and still get you there safely.

Constellations of satellites such as these can certainly assist with navigational problems around a planet. However, when moving beyond the planet some other means of navigation must be found. The limiting factor is simply one of power, and it's not hard to see why. The Sun, for example, radiates energy at the rate of about 30 million billion billion (10^{25}) watts. By the time the Sun's energy reaches the Earth's atmosphere, the solar energy per unit area is about 127 W/ft^2. In comparison, the GPS satellites bombard the Earth with signals of about 0.000,000,000,000,000,4 (10^{-16}) W/ft^2. That's about the same energy per unit area that a 25-watt lightbulb in New York will produce when the light arrives in Los Angeles (excluding the effects of smog). Although small, the amount of energy these satellites broadcast is large enough to be detected by instrumentation on the Earth. If we try to send the same signals from Earth to Mars, however, the energy flux received from the satellites would have fallen to less than one-billionth of the value on the Earth's surface. To still be detectable on Mars, we would have to increase the signal

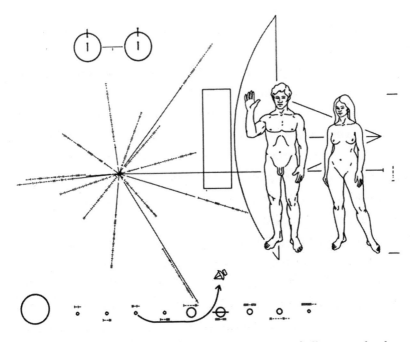

Figure 46. The plaque attached to the *Pioneer 10* spacecraft illustrates the shape of its human creators, our size relative to the spacecraft, the spacecraft's origin—the third rock from the Sun—and the Sun's position relative to several naturally occurring radio sources called pulsars. (Courtesy NASA)

strength from the satellites by a factor of one billion. Even if we do this, all of the GPS satellite signals would appear to come from the small dot in the sky called Earth and would not yield any significant navigational information, save as a homing beacon. In spite of modern advances, navigating by the stars—the tried and true method of ancient mariners—remains the best possibility for future star travelers. The plaque attached to the *Pioneer 10* spacecraft (fig. 46) shows Earth's location relative to a number of special stars called pulsars. By triangulating from the pulsars identified on the *Pioneer* plaque, future alien civilizations should be able to locate the star (our Sun) from which the spacecraft came. Unfortunately, all of the effects of the space environment would probably degrade the plaque into an unreadable blob long before *Pioneer* completes its multimillion-year journey to the next star.

Although space exploration has more possibilities of uniting mankind across cultural barriers than any other endeavor, the cost of sending people into space is prohibitive. The true cost of a single Space Shuttle launch is about $1 billion. The cost to assemble the International Space Station will probably reach $80 billion, plus a maintenance cost of an additional $2 billion per year. In comparison, unmanned spacecraft can explore our solar system far more cheaply. Scientific probes like Galileo, Cassini, Voyager, and Pioneer cost on the order of $1 billion to build *and* launch and generate far more data—for far longer periods of time—than could a week-long Shuttle mission. The GPS navigational satellites cost about $30 million each, and can operate for 10–15 years. This cost difference alone accounts for the fact that the United States and Russia average about ten manned space missions per year, while worldwide there are well over one hundred unmanned satellite launches per year. The long-term exploration of space will continue to place a great emphasis on unmanned spacecraft, which are cheaper, last longer, and can be sent into more hostile environments than can spacecraft carrying a crew. Much of the renewed interest in space in recent years has been fueled by business interests, driven by the realization that some things can be done more cost-effectively by satellites. As the cost of building and launching satellites continues to drop, it may well be that the long-term exploration of interplanetary or interstellar space will depend not on governments, but on private investors.

The Shape of Things to Come

One can't discover new lands without consenting to lose sight of the shore for a very long time.

As Andre Gide, the Nobel Prize–winning author, indicated over half a century ago, to gain a clearer picture of what the future may hold we must first let go of our vision of the past. In the twentieth century, space exploration began as a competition between nations to prove technological superiority and then evolved into a cooperative effort aimed at yielding a better understanding of the world around us. In the twenty-first century, the exploration of space may be driven by thriving industries that today are in their infancy.

For example, materials-processing could be a viable use of space. Inside the heart of most modern electrical devices is a computer chip of some sort. These chips are all manufactured in ultra-clean facilities on the ground because their small sizes (microinches) require them to be made of pure materials. The presence of even the smallest amount of dirt in its interior can render a computer chip useless. There is enough "dirt" floating in the air of a typical room to deposit significant amounts of dust on surfaces during a single day, as anyone who wears glasses knows. If dust from the air settles onto computer chips, the dust particles will interfere with the electrical circuits printed on the chips. In the weightless environment of space, dust would not settle and computer chips would not be harmed by it.

Another limiting factor in the manufacture of computer chips is gravity itself. Many computer chips are made from silicon crystals. These crystals can be grown in the laboratory, but their size is limited because their own weight will eventually prohibit further growth. If these chips could be grown in space, where they would be weightless, we should be able to grow larger chips more effectively. Experiments in weightlessness, on Shuttle missions, have shown that the idea works, but at the moment the cost of getting to space is so great that we can't afford to build large numbers of computer chips there—yet. In the future, this area of technology will undoubtedly continue to be explored.

Mining is another definite possibility associated with long-term space exploration. Unlike mining here on Earth, space miners will not be immediately interested in finding precious gems, minerals, or oil. They'll be interested in those items that are of most critical importance to a spacecraft and its crew—air and water. It takes about 60 pounds of water and 2 pounds of oxygen per person per day just to keep a crew alive on a long mission. (This may seem like a lot of water, but water is used not only to drink, but to bathe, wash clothes, and so on.) If we can mine oxygen and water from the Moon, or from other planets, we will be able to restock our supplies en route. This will leave more room on the spacecraft for other things besides life-support systems. It simply makes sense to follow the Boy Scout's example and try to live off the land. When you drive to visit your cousin, you don't have to take along enough food and fuel for the round-trip journey because you know you can purchase both along the way. Knowing that you don't have to take along such items

Figure 47. Artist's conception of the International Space Station after its completion in 2004. (Courtesy NASA)

means you can get by with a smaller, less costly, vehicle, or fill up the larger one with children's toys to combat the "are we there yet" problem. If we could travel to the Moon or another planet and refuel once we get there, we'll be able to take more data-gathering, or revenue-generating, equipment along.

The last and final sign that the space age is truly upon us will surely be the emergence of a thriving space-tourism business, an industry whose sole purpose will be to provide a means for average people to get to space. Who wouldn't want to visit a space station during summer vacation, or spend a little quiet time on the dark side of the Moon? Each year, 10 to 12 million people visit the U.S. Air and Space Museum, the Kennedy Space Center launch site, space camps, and related destinations. The terrestrial space-tourism business probably generates revenues of about one billion dollars per year. Market studies done by the National Aerospace Laboratory in

Japan indicate that space travel and tourism could easily generate annual revenues of as much as ten billion dollars per year. Even today, the "X Prize" Foundation is hoping to stimulate the development of commercial space tourism by offering a ten million dollar prize to the first private team to build and fly a reusable spaceship capable of carrying three individuals on a brief 15-minute flight to space. The owner of a well-known hotel chain recently announced a commitment to develop the design for a luxury hotel in space. Other investors have promised to save the aging Russian Mir Space Station and convert it into an orbiting hotel. Such designs won't come cheaply. It will take billions of dollars to fund the team of developers needed to transfer the concepts into credible designs that are both buildable and affordable.

Perhaps the "final" use of the final frontier will actually be as a burial ground. For the past several years, it has been possible to launch the ashes of your dearly departed into the great beyond for less than five thousand dollars per person. Some notable participants include 1960s drug guru Timothy Leary and *Star Trek*'s creator Gene Roddenberry. Comet hunter Gene Shoemaker, of Shoemaker-Levy fame, was rewarded with an even more unique ride when some of his ashes were carried aboard the *Lunar Prospector* spacecraft. He will have the honor of becoming the first person to become interred on the Moon when the satellite runs out of fuel and impacts the lunar surface.

Undoubtedly, many more uses of space will be found in the next few years. After all, who would have predicted the appearance of wonders like the Internet twenty years ago. A critical factor in many uses is simply the cost: How expensive will it be to do something up there compared to down here, and are we willing to pay that price?

FOR FURTHER READING

Nontechnical Books

Chaikin, A. *A Man on the Moon*. New York: Penguin Putnam, 1994.

Collins, M. *Flying to the Moon: An Astronaut's Story*. 2nd ed. New York: Sunburst, 1994.

Davies, J. K. *Space Exploration*. New York: W. & R. Chambers, 1992.

DeVorkin, D. H. *Science with a Vengeance: How the Military Created the U.S. Space Sciences after World War II*. New York: Springer-Verlag, 1993.

DeVorkin, D. H. *Race to the Stratosphere: Manned Scientific Ballooning in America*. New York: Springer-Verlag, 1989.

Hickam, H. *Rocket Boys*. New York: Bantam Doubleday Dell, 1998.

Joels, K. M., Larkin, D., and Kennedy, G. P. *The Space Shuttle Operator's Manual*. New York: Ballantine Books, 1988.

Kranz, G. *Failure Is Not an Option*. New York: Simon and Schuster, 2000.

Krauss, L. M. *The Physics of Star Trek*. New York: HarperCollins, 1996.

Lattimer, R. *All We Did Was Fly to the Moon*. Gainesville, Fla.: Whispering Eagle Press, 1995.

Lovell, J., and Kruger, J. *Apollo 13*. New York: Simon and Schuster, 1995.

Neal, V., Lewis, C. S., and Winter, F. H. *Spaceflight—A Smithsonian Guide*. New York: Macmillan, 1995.

Still, R. *Relics of the Space Race*. 2nd ed. Roswell, Ga.: PR Products, 1998.

Zubrin, R., and Wagner, R. *The Case for Mars*. New York: Touchstone, 1996.

Technical Books

Larson, W. J., and Pranke, L. K., eds. *Human Space Flight Analysis and Design*. Dordrecht, The Netherlands: Kluwer, 1999.

Sellers, J. J. *Understanding Space—An Introduction to Astronautics*. New York: McGraw-Hill, 1994.

Tribble, A. C. *The Space Environment—Implications for Spacecraft Design*. Princeton, N.J.: Princeton University Press, 1995.

Wertz, J. R., and Larson, W. J., eds. *Space Mission Analysis and Design*. 3rd ed. Dordrecht, The Netherlands: Kluwer, 1999.

Internet

AUTHOR'S NOTE. Because information on the Internet changes so rapidly, it is difficult to give references that will remain valid for some time. Please consult my home page at *www.alantribble.com*, where I will try to maintain some links to sites of interest.

INDEX